The Cyber War is Here

The Cyber War is Here
U.S. and Global Infrastructure Under Attack: A CISO's Perspective

Marc Crudgington

All Rights Reserved. No portion of this book may be reproduced, stored in a retrieval system, or transmitted in any form or by any means – electronic, mechanical, photocopy, recording, scanning, or other – except for brief quotations in critical reviews or articles without the prior permission of the author.

Published by Game Changer Publishing

Paperback ISBN: 978-1-962656-47-4
Hardcover ISBN: 978-1-962656-48-1
Digital: ISBN: 978-1-962656-49-8

www.GameChangerPublishing.com

DEDICATION

To my supportive and amazing wife, Maricris, and two sons, Jake and Luke, who have always encouraged me with praise, love, and gratitude. You deserve way more than I could ever give, but I will always do my best to give all of me to each of you.

To all my peers, thank you for being some of the best in the world at what you do and making cyber fun, thought-provoking, and gratifying every day. To my friends and family, thank you for being part of my immediate family's and my life; your generosity, love, and support have always been endless.

To everyone who is reading this book and perhaps taking their first step or encouraging others in the field of cybersecurity: Jump in! It is a vast field of opportunity with many disciplines to enjoy. Whether you realize it or not, we are all targets of cyberattacks, every person and every business. Cybersecurity is all our responsibility.

I acknowledge and thank all of you. Godspeed on your life's journey.

Read This First

Just to say thanks for buying and reading my book, I would like to provide a way for you to connect with me, no strings attached!

Simply Scan the QR Code Here:

The Cyber War is Here

*U.S. and Global Infrastructure Under Attack:
A CISO's Perspective*

Marc Crudgington

www.GameChangerPublishing.com

Confessions and Acknowledgements

I'd like to start by thanking those who purchased my first book, *The Coming Cyber War: What Executives, the Board, and You Should Know*. I am profoundly grateful for all of the compliments, interest, talks, and ratings the book has generated. I've heard from people with little to no experience in cybersecurity to CISOs (Chief Information Security Officers) at Fortune 500 companies and board members of large public companies, thanking me for the content and raising awareness of the cybersecurity risks we face.

The hidden joke of the book is in the title, *The Coming Cyber War*. Yes, spoiler alert: we are in a global cyber war. Some may not think that, and many do not know that we are. Every day, we read major headlines about nation-state attacks, whether against the United States or global ally government institutions or against some of the largest companies in the United States and around the world. The news is endless.

I never imagined I would be writing a second book, especially after figuring out why people do not write a first book, because you are actually WRITING A BOOK! Though rewarding, it is tedious, and there is a bit of apprehension about what others might think. Not to mention the time and opportunity costs that go by the wayside. However, I feel like many highly publicized large-scale cybersecurity breaches and events occurred shortly after the first book was published, and those events warranted writing another. The last three years, from October 2020 until October 2023, have been nothing short of complete disruption and chaos—the so-called "wake up" call. It seems like we've had those just about every quarter, from SolarWinds to Colonial Pipeline to Microsoft to Clorox to MGM to large social media companies, municipalities, and national government institutions.

I believe authoring a book on why we, both the United States and our global allies—citizens and government—need to elevate our cybersecurity resilience acumen and knowledge of cyber warfare is a matter of national security as well as economic security. This book is for anyone who is interested in understanding why topics related to cybersecurity matter on a global basis, especially to America's critical infrastructure.

Right now, we've come to a crossroads. Many national directives, directives from the President of the United States of America, throughout administrations from Ronald Reagan and George H.W. Bush to the current administration of President Joe Biden, have been released related to cyber warfare. Each administration had a concern about cybersecurity and the security of America and our allies related to potential cyber risk. In fact, in 1982, Ronald Reagan authorized one of the first initial cyber-related attacks on the Soviet Union, which was against a Siberian natural gas pipeline. Many believe that Stuxnet was the first cyber weapon. However, some "code work" was done related to the operations of industrial control systems used to operate a Siberian pipeline, which caused a catastrophic explosion, essentially vaporizing a large section of the pipeline. No one was injured, but those in the profession who know understand that was a foretelling of where we're at today.

Numerous cyberattacks and cyber warfare are occurring that your average American, or from any other country, does not fully know the significance of what is transpiring. Nor do many fully comprehend why the state of cyberspace, whether it is personal cyberattacks, intellectual property theft, identity theft, or large-scale attacks and cyber warfare that make the news impacts them. Whether it is an attack at their place of business, within their home and their own privacy data, or their children's or grandchildren's privacy data, if you just listen to the news, you may think that cyberattacks only occur every few months. Nothing can be further from the truth. Cyberattacks happen on average every 39 seconds.

You may be wondering about how I got here. I'd certainly encourage you to read the Robert Frost poem "The Road Not Taken." My background is approximately 30 years in cyber and technology, going back to 1992. I started off my career in the U.S. Air Force. Interestingly enough, cyber was not what we know cyber to be today. However, threat intelligence existed, and monitoring the actions of other countries through electronic means, spying, and espionage has been around for decades. Cyberattacks were not often mentioned, if at all. I spent time in threat intelligence and a number of other technology disciplines while in the Air Force.

Once I got out of the Air Force, I went to work in industry, spent roughly ten years in Silicon Valley at various technology companies, doing security work as well as other information technology work, and then became an executive at a company in Southern California, Orange County, after spending a few years in Big Four consulting. While in Orange County, I got my MBA at the University of California, Irvine, Paul Merage School of Business. After living in Southern California for about eight years, I moved back to the Houston, Texas, area, where I worked as Chief Information Security Officer (CISO) for a large bank headquartered in

The Woodlands, Texas, and then on to my current position as a CISO at a large public company.

I have several board-related, cybersecurity, technology, and business certifications. I have authored several articles on cybersecurity and technology, spoken at numerous conferences and for companies on cybersecurity, and currently hold or have held several advisory board positions related to providing technology and cybersecurity advice. I have owned a cybersecurity solutions provider and consulting company, which I started in 2018, where I performed duties as a virtual chief information officer (vCIO) and a virtual chief information security officer (vCISO). Currently, I am on my third opportunity as a full-time CISO (Chief Information Security Officer). I had the privilege to attend the FBI's CISO (Chief Information Security Officer) Academy in February 2017. I have won a national award for cybersecurity executives, keynoted several conferences, and served on many panels. My experience is vast in both cybersecurity and technology. The first book I wrote is called *The Coming Cyber War: What Executives, The Board, And You Should Know*.

Foreword by Clif Triplett

Cybersecurity risk management has never been more important than it is today. Understanding the dynamic nature of cybersecurity has become essential for all those engaged in the operation and protection of the nation's critical infrastructure. The sophistication of the cyber threat has been progressing at an alarming rate, and staying abreast of it has become a challenge for anyone who needs to engage daily in the cyber war.

The Cyber War is Here provides the reader with a very efficient way to understand the global situation better. The array of topics addressed is significant and succinct. The book's objective is to provide the reader with the information they need to understand the threat from various perspectives, but not necessarily the details of a specific technical threat response. The primary audience is senior management, but the daily cyber warrior's educational value can be equal. Embracing the content of this book can allow anyone to be far more conversant on the global cyber threat, its potential impact, and, unfortunately, being able to discuss a frightening look forward to the risks ahead.

If you, the reader, are a Chief Information Officer (CIO), Chief Information Security Officer (CISO), risk manager, or consultant, this book can improve your understanding of the threat and put it in the context of recent history. It will allow the reader to be more comfortable discussing this with their C-suite peers, the board, or, as consultants, their potential clients. *The Cyber War is Here* will provide actionable information, enabling impactful dialog with peers and providing valuable education for the board of directors.

Nation-states and criminals have found cyberattacks to be a weapon or tool of choice in their arsenal. International prosecution of criminals is less than 1 percent by most accounts, yet the revenues they realize are exponentially growing. Nation-states like cyber weapons because the impact of the required investment is superior to conventional weapons, and confidence in the attribution of the attack is very often ambiguous. Cyberattack success has motivated nation-states and criminals to invest even more significant amounts into their efforts.

With the advent of artificial intelligence (AI) entering the fray of cyber weapons, the rate of sophistication in cyber weapons will take another significant leap forward. The current political unrest in Russia, Ukraine, Iran, North Korea, and China has also encouraged more significant nation-state investment levels. We live in a connected world where our lives depend highly on infrastructure operating on shared networks utilizing software and software-defined devices. Governments have recognized the challenges facing their countries.

The regulatory environment everywhere is trying to keep up. Regulatory bodies across the globe are generating new guidance several times a year. The legislation they enact can significantly impact businesses and is usually trailing behind the progression of the threat. These regulatory bodies have moved beyond the national government, but now we see increased regulations coming from the States and provinces. How do we keep up?

It is not a choice, and it is not easy. *The Cyber War is Here* is one of the best tools to improve your understanding of today's cyber climate. As the book is titled, the cyber war is here and ongoing. Whether you serve in the government or private industry, your organization is being attacked and interrogated for weakness. Will you be ready when your vulnerability is discovered and exploited?

Failure to recognize the reality of the threat and begin to take remediating actions now can have severe consequences for those in charge and the enterprise as a whole.

Whether you read every word or use it to reference a specific topic, this book should be in your library.

— Clif Triplett, Executive Director Cybersecurity & Risk Management, Kearney, former advisor to the Office of the President of the United States, Presidential Executive Fellow for Cybersecurity, U.S. Military Academy West Point graduate

Table of Contents

Introduction 1

Chapter 1 – Crisis Among Us, What is Leading Us to Here 11

Chapter 2 – National Security and Systemic Risks 27

Chapter 3 – Cyber Threat Actor Landscape and Motives 49

Chapter 4 – Threat Actor Tactics, Techniques, and Procedures 71

Chapter 5 – Mitigating Risks and Hacking Back 97

Chapter 6 – The Board of Directors and CISO Collaboration 141

Chapter 7 – The Future Is Here 155

Chapter 8 – America's and our Allies' Path Forward 171

Conclusion – Is a Cyber War Really Here? 177

Bibliography 183

Introduction

I'm writing the book for anyone interested in learning more about cybersecurity and cyber warfare and why they matter globally, especially in America. I think right now, we've come to a crossroads. You may be hearing a lot about national directives, directives from the President, that cross administrations from the Reagan and Bush administrations to the current administration of President Biden. There was and is a concern for cybersecurity. In fact, as stated, President Ronald Reagan authorized one of the first initial cyberattacks related to a pipeline in Siberia. Many people believe that Stuxnet was the first cyber weapon.

However, some coding was done on industrial control systems, and it ended up blowing up a Siberian pipeline. No one was injured, but those in the profession who know feel that that was the catalyst for really getting us to where we are today. I think there are so many things that are going on in cyberspace and cyber warfare that individuals that are your average American or from any other country just don't know the significance of several things that are going on and why this impacts them; one, at their place of business, as well as within their home and their own privacy data or their kid's privacy data.

Over the last few years, from early 2021 and throughout the Ukraine and Russian War, we've seen that cybersecurity has received more attention due to various newsworthy incidents. But there were several attacks in the past, obviously, that we're all aware of. I mentioned Stuxnet, but other personal Home Depot banking, Colonial Pipeline, Solar Winds, T Mobile, government agencies, you name it. There have been many critical cybersecurity attacks on our infrastructure and our allies' infrastructure. It is affecting us in more ways than individuals know. It affects us from an intellectual property perspective, affects us financially, and affects our jobs. Obviously, the expenditures that we have to put out to defend our nation against cyberattacks and defend our businesses are costing America and other countries globally a lot of capital expenses. We do have a shortage right now in the U.S. regarding cybersecurity professionals. There are predictions that well over 3.5 million job openings in cybersecurity will go unfilled in 2023. Cybersecurity has negative unemployment in 2023—more jobs than there are people interested and qualified to fill them. Given that, this book is also for those thinking of a cybersecurity career. But it's really written not only for the average American or global citizen but also for executives in business, explaining why they need to pay attention and protect their company's assets.

There are several legislation introductions that impact boards, company executives, and officers that are on the horizon, as well as legislation that impacts critical infrastructure companies. There are 16 critical infrastructure domains. And I would guess that almost a very large percentage of American companies fall into the critical infrastructure category. Now, the government's not necessarily worried about every single company, just the major ones. And we saw that with Colonial Pipelines, SolarWinds, and those that would have a material effect on America. This book can serve as an educational rallying cry for us to get involved in cybersecurity and raise our level of awareness. I often state that cybersecurity and cyber warfare are very much like the ocean's current;

you don't know that they exist until you're actually in them. Once you are, it can be fairly devastating, especially during a cyber crisis.

For those individuals who have had to experience an attack at their company, it can be pretty daunting and emotional to resolve. Even if it is just a minor cyber incident, it can quickly get out of hand. That is really who I believe the book is for. Your average American doesn't need to be a cyber expert, but also business executives and the board of directors for companies around America and the globe. My hope for this book is that when individuals read it, they will sense it to be educational and useful in, at a minimum, capturing their attention to examine the topic a little further. If they should read it and feel they have become aware of what is going on across the globe and with America, especially why we should have certain cyber laws, not only data privacy but a cyber peace treaty, then even better. Certainly, cyber is used for espionage and spying, and we understand that. I think we're going to cross a critical red line when individuals start getting physically hurt, or death occurs because of cyberattacks.

Those types of events will push us further into a state of war, both in cyberspace and in physical space as well. And that may just trigger something that, who knows, could trigger the next World War when it comes to a cyberattack that injures or causes death on a broad scale. There will be many scenarios for how that could play out. I believe what individuals will learn from this book is the history of cyber warfare and many things that have happened in the past, especially in the last three years since publishing my last book. There has been an increasing escalation in attacks, cyber warfare, and the technology used since late 2020. We have seen a significant uptick in attacks on critical infrastructure and major attacks that have impacted individuals across all sectors in numerous countries, from government agencies to school districts to personal data to company data to a very big attack that had an enormous impact on America, which was the Colonial Pipeline attack. If you are a board member, an executive at a company, an individual contributor who has access to private data or data that might be company confidential or intellectual property, or a novice on the topic, this book is right for you. It has purposefully not been written in highly technical jargon.

One of the things that we talk about in cybersecurity amongst peers is that when a significant cyber incident occurs, one of the first things you hear is, "Well, this is a wake-up call. This is a wake-up call." When are we actually going to wake up? I think it is time to realize and admit that we are in a cyber crisis. The cyber war is here. We all should increase our awareness of cybersecurity and realize its impact on each of us. Cyber warfare, that is. It is a systemic threat. Everything from the actual war that we see going on in Ukraine to the typical cyberattack on our personal data, attacks on our companies, and extorting us and our data through a ransomware attack. Essentially, at any moment, there's a cybersecurity crisis going on across America and with our allies; individuals may not know the scale to which attacks are striking.

Dwell time, which is the time when a threat actor gets into a company's infrastructure and stays there, pillaging data, pivoting across the environment, and plotting where they should move next within your infrastructure, can be weeks, even upwards of 200+ days, depending on the industry you're in. As you read this, you or someone close to you may be experiencing a cyberattack and don't even know it.

I mentioned, from an intellectual property perspective, that this is costing companies billions of dollars. China, among other countries, is able to steal our intellectual property and then duplicate products for their own benefit; our commerce infrastructure is under attack. Many have heard the news of intellectual property theft across numerous industries. There are several stories about automakers having data stolen and Chinese clone cars that take a door off of one of our cars made in America, and that exact door or part will fit. Some of the automobiles they're making over there are almost exact replicas of cars made in America or other countries.

A good example would be the Range Rover Evoque. Now, China has a car they call the Landwind X7. It looks almost exactly like the Range Rover vehicle. That's happening across all industries, not just the auto industry. Technology, manufacturing, pharmaceuticals, energy, and entertainment industries are not left out, and it is not stopping. Our personal data and identity are at high risk of being stolen by any threat groups representing different countries or threat actors from those countries.

A bit of information on why I believe this book is important, especially this topic, as well as a little about my background. I do hope you get a lot out of reading the book, and it is written at a level for the novice and expert to get value from. Again, it is purposely not written to be too deep in cyber or tech jargon. I firmly believe this is a topic that is top of mind for many executives, business leaders, company board members, and those in government agencies.

I do hope, whether you are knowledgeable about cybersecurity or a novice, that you enjoy reading the book.

A WORD ON THE MGM AND OTHER RECENT BREACHES (2023)

You all know the saying, "What happens in Vegas, stays in Vegas." Well, this didn't. The details of the MGM breach have been equally devastating and mind-boggling. The MGM cybersecurity breach that was disclosed in early September 2023, along with some of the other breaches that occurred in that timeframe, Clorox and Caesars, have been catastrophic in nature and are yet another "wake-up call" for Boards, business executives, cybersecurity, and the public. The Caesar's breach that was disclosed in September 2023 resulted in over 41,000 Caesar's customers having their personal data stolen. Caesars was additionally dealing with ransomware from the same threat group, Scattered Spider, that successfully attacked MGM. Caesars was able to negotiate a ransomware payment from $30 million to $15 million. With the exception of stolen data and a little disruption, Caesars was able to operate as normal. Clorox, on the other hand, is still suffering from the breach that has affected them – everything from operational impacts causing shortages with their products to devastating financial losses in revenue and shareholder value. Clorox announced in early October that the breach would cost them between $487 million to $593 million in a net sales decrease, or 28% - 23% from the year-ago quarter. Organic sales are expected to decrease between 26% to 21% for the quarter, compared to mid-single-digit growth. The breach is affecting diluted and adjusted EPS – earnings per share, with significant losses, and their stock price has suffered greatly.

The MGM breach has been just as devastating as Clorox, but maybe more so, as it was going on while thousands of casino patrons were visiting any one of their many casinos. It broke nationally and globally. There were wide-scale outages from hotel room keys to slot machines to guest registration to dining operations, websites, and more. The breach was so dire at one point MGM had to result in handwritten, yes, you read that right, handwritten receipts for casino winnings and went into manual mode, according to an MGM spokesperson; imagine explaining that for a global casino operator. A substantial amount of guest data, including names, addresses, phone numbers, email addresses, and, most impactful, passport numbers, were stolen. The threat group, Scattered Spider, even posted on their dark web blog, ALPHV, their reasoning for the breach and follow-up activities.

What caused the breach is quite extensive, and most are still yet to be dissected. What has been widely reported is that Scattered Spider trolled LinkedIn to find an individual who might have high-level IT administrator privileges. They then spoofed (pretended to be) that employee's phone number and called the MGM help desk/technology support team, masquerading as that individual and asking their MFA (multi-factor authentication) login credentials to be reset – a vishing attack. Within 10 minutes, they were inside MGM's infrastructure. They then proceeded to compromise a super user of an IT system related to managing employee access and identity, among other technology services. Once in this system, they compromised numerous other highly privileged IT administrator accounts. After this was done, they essentially could control the technology infrastructure at MGM. They spent the next four months exfiltrating/taking sensitive data from MGM's network. They then proceeded to encrypt many of MGM's servers and disrupt the services that run on those servers. This explanation leaves out a lot of detail, as much is still unknown, but the key points are they were able to call the IT help desk to penetrate MGM infrastructure in a very easy fashion and then spend four months in the infrastructure relatively undetected. Frightening, to say the least.

There are many consequences for MGM and lessons to be learned. Consequences for MGM include the financial losses MGM is suffering and MGM shareholders. Reputation damage – this is not the first time MGM was breached, and the way MGM was breached, this time raises a lot of questions. Operation disruption and opportunity cost that MGM has incurred or will incur. Strengthening MGM's cybersecurity infrastructure will be substantial; it needs to be evaluated from top to bottom across all aspects of building a cyber-resilient program. Executive damage – Scattered Spider posted information about executive stock sales that will raise questions. Lastly, the legal implications will most certainly be many. Customers have already filed lawsuits. Regulatory bodies governing MGM will definitely impose any fines for not protecting data. It will be interesting to see this from a state and national lens.

The MGM breach serves as a stark reminder of the importance of cybersecurity to all organizations worldwide. Immediate lessons learned can be related to user awareness training at all levels of organizations; user authentication methods when someone calls the help desk (was the employee number – something only the employee and very few would know – required); configuration of the IdP (Identity Provider) system – was it configured with recommended best practices, when was the last time a health check was done; how strong was MGM's continuous monitoring program related to anomalous activity; where was the system that protected and/or managed privileged accounts – these can be configured in a manner that would have set off many red flags related to password changes and anomalous activity; was their segmentation in the MGM infrastructure that could have made it more difficult to pivot around the infrastructure; obviously systems with sensitive/customer data in them should be considered Tier 0 critical assets – how were they protected – dual control, service tickets and alerts for when accessed, encrypted data with only a limited number of people able to do full decryption that also requires dual control; endpoint security systems – how robust were they; and cybersecurity empowerment – was the cyber staff, from analyst to CISO, ignored or risk accepted if they brought issues up, were they looked at enablers of the business or sand in the gears, how was risk in the organization viewed – finger pointing or let's team together to mitigate; and many other lessons can be learned from what is unknown.

These breaches are sobering reminders that no organization, regardless of industry, size, or location, is immune to a cyberattack. The breaches exposed weaknesses in cybersecurity measures that can be mitigated – yes, some may take an elevated effort, but others are foundational cyber hygiene. They have/will result in severe consequences for the companies, both financially and reputationally. These breaches are just a few of the recent ones; each highlights the need for all organizations to continuously prioritize and invest in cybersecurity, stay vigilant against evolving cyber threats, and view cybersecurity as a business enabler – like the brakes on a Ferrari – they are not in place so you can stop, but so you can drive 160 miles per hour safely. How many CEOs, Board members, experienced government officials, and cyber experts need to warn of the risks of not taking action? With advancements in technology that can be used for good and bad, it is only through proactive, design to be cyber resilient efforts that companies can hope to avoid becoming the next victim in the ever-escalating cyber war. Godspeed to all those who protect our nation and our allies against cyber threats.

CHAPTER 1

Crisis Among Us, What is Leading Us to Here

In today's interconnected world, cybersecurity has emerged as a critical concern affecting individuals, organizations, and nations alike. The United States, being at the forefront of technological advancement, faces numerous challenges and significant security threats in our digital landscape and our critical infrastructure. And this goes for countries across the globe. There are very profound effects of cybersecurity and cyber threats on Americans and citizens around the world. Some of those are related to identity theft, job losses, systemic risks, and alarming issues around children's identities being stolen by people who don't realize it. Then, when a child turns 18, the threat actors know this and start exploiting the child's identity, and these young adults start seeing the effects on their credit. These are all major concerns for Americans.

Identity theft is a very prevalent cyber threat that wreaks havoc on individuals and our economy. Our cybercriminals utilize various tactics to gain unauthorized access to our personal information, leading to financial and emotional distress, along with the headaches of trying to clear or reestablish your identity. Americans, especially, given we are one of the richest countries, if not the richest country in the world, have experienced the debilitating consequences of having our identity stolen, including compromised financial records and accounts, damaged credit scores, and then rebuilding.

This not only affects the individual's well-being but also strains the resources of our financial institutions and law enforcement agencies, leading to significant economic losses, time losses, and job losses. Cyber threats pose a substantial risk to businesses, resulting in job losses for American workers. Many people don't realize this. But when an organization falls victim to a cyberattack, it can shudder, especially if it's a small business. They often face significant financial setbacks in trying to recover from the impact of a cyberattack. It may take days, weeks, or months to recover. There's reputational damage due to a cyberattack and disruption of operations across the company.

In response, many companies may resort to downsizing, layoffs, or even bankruptcy. And there are numerous occasions where companies have had to shut down. The result is that employees are rendered unemployed or struggle to find new employment opportunities. And if you're on the cybersecurity team, it can really affect you long-term if you're at a company with a major breach. Moreover, the fear of cyber threats can lead to reduced innovation and investment, hindering our country's economic growth and employment prospects for all Americans.

Systemic risks are omnipresent. Cyber threats extend beyond individual and corporate entities, presenting a systemic risk to critical infrastructure, government institutions, and national security. Attacks that target vital sectors of our critical infrastructure, such as energy, finance, health care, transportation, and education systems, can have far-reaching consequences. They disrupt essential services, compromise public safety, and undermine even further our trust in government institutions, whether they're local, state, or national.

The interconnectedness of our modern and technological systems amplifies the potential for cascading failures, highlighting the urgency of robust cybersecurity measures and cooperation among public and private entities. We have some great examples of this, with the SolarWinds breach that happened at the end of 2020, as well as the Colonial Pipeline attack that shut down the pipeline across the Eastern Seaboard. Those impacts were felt far beyond the direct impact.

Now, children's identities are becoming one of the scariest types of attacks. It is an alarming trend in the realm of cybersecurity when a child's identity is stolen. They may not know it for years. Parents may not know it either. Even the entity where the data was stolen may not know it. Children have a clean credit history, making them a prime target for cybercriminals seeking to exploit their personal information for illicit purposes. Stolen identities can be used to open fraudulent accounts, commit financial fraud, or engage in other criminal activities, causing significant harm to a child's future prospects. Moreover, the long-term repercussions of children's identities being stolen are often overlooked, as rectifying the damage can be lengthy and complex to recover from.

Cybersecurity and cyber threats have a profound impact on Americans across multiple fronts. Identity theft inflicts financial and emotional distress on individuals. The job losses can be significant. Addressing these challenges requires a multifaceted approach involving collaboration between government, industry, and individuals. All Americans and citizens across the globe should heed warnings and pay attention to these impacts. The increased investment in cybersecurity infrastructure and public awareness campaigns that you now see as part of the Department of Homeland Security, CISA, and FBI are very important. Legislation at a state and national level and international cooperation are imperative to safeguarding the digital landscape and protecting Americans and our allies across the globe from the pervasive threats posed by cybercrime. Only through collective efforts between government agencies, state agencies, national agencies, companies, and citizens can we build a resilient and secure cyber ecosystem for all of us to benefit from.

In the digital age, cyber warfare has transcended traditional battlefields, extending into the realms of cyberspace. Cyber warfare, the use of computer technology to disrupt or destroy an adversary's infrastructure, has emerged over the last 10+ years as a potent weapon in the modern era. The history of cyber warfare, exploring several things that have happened in significant cyberattacks and their far-reaching implications, focuses on four major cyberattacks that have gotten us to this point and have shaped the landscape of cyber warfare. Those are Stuxnet, Colonial Pipeline, SolarWinds, and the Ukrainian state nuclear tanks during the Russian-Ukraine conflict. The origins of cyber warfare can be traced back to the very early days of computing when nations began to realize the potential of utilizing information technology in warfare.

As the internet became increasingly intertwined with critical infrastructure, vulnerabilities continued to grow rampantly, and the opportunities for nation-states and other criminals or those with nefarious intentions could exploit those. And the late 20th century is really the dawn, or when nations began developing offensive cyber capabilities to gain an edge over other countries or in times of conflict.

We saw back in 1982, though, that there was an instance where the U.S. corroborated with our allies, mainly Canadians, to blow up a Siberian pipeline. And this was obviously when the USSR—the Soviet Union—

existed; they needed to update their critical infrastructure. The Soviets came to the United States, and obviously, we said, "No, we're not going to help you." They then asked our allies in Canada for assistance. The United States C.I.A. tipped off the Canadians. In collaboration, the Canadian government passed the Soviets a code that would cause an accident. This incident or this occurrence was authorized by Ronald Reagan, and it was a plan to obviously sabotage the pipeline, causing catastrophic disruption to services. One of the other prevalent cybersecurity attacks that sparked warfare was Stuxnet, a significant cyberattack on Iran and its nuclear program.

Stuxnet was a malware worm that would travel through an organization's infrastructure. It targeted Iran's nuclear program, specifically the use of uranium enrichment facilities. We'll dive into this a little bit more later on. The operation was a joint effort between the U.S. and Israel to cause Iranian nuclear centrifuges to essentially freeze up due to spinning at a significantly high rate—way out of the normal operating range. The infiltration went undetected for some time.

We witnessed the SolarWinds breach, which was a supply chain attack in late 2020 that really showed us how cyber warfare has evolved. The attack was within the SolarWinds software supply chain, specifically the SolarWinds Orion platform, which is widely used in government agencies as well as major corporations across the globe. It was suspected to be the nation-state hacker group known as Nobelium, believed to be from Russia, and the vulnerability allowed them to get unauthorized access to numerous organizations and entities. The vulnerability existed for some time as it was injected into the platform early in 2020, and updates were provided to customers with the malicious code on the update platform.

The Colonial Pipeline ransomware attack, which occurred in 2021, was another notable cyberattack that was a 'wake up' call. The threat group committing this attack was a Russian criminal gang called DarkSide, and it was against the Colonial Pipeline I.T. network and systems. This attack shut down the pipeline due to worries the attack may spread into the critical infrastructure. The incident caused widespread fuel shortages across the U.S. and had far-reaching ramifications for our critical infrastructure. Multiple government agencies either assisted in response and/or were notified of the incident, including the FBI, the Cybersecurity and Infrastructure Security Agency, the U.S. Department of Energy, and the Department of Homeland Security. Numerous legislative mandates were enacted.

Most recently, we have seen cyber warfare during the Ukrainian and Russian conflict, where Ukraine has experienced several attacks on its critical infrastructure. Even going back to 2015 and 2016, Ukraine suffered multiple cyberattacks on its power grid that were state-sponsored by Russia before the Crimea attacks. These types of attacks can be a prelude to war and blur the line between cyber and kinetic warfare. Many have the potential to cause physical destruction and loss of life.

We have had many attacks on our critical infrastructure across the globe before these events. They have become increasingly widespread in the last few years. Other major attacks have included a 2012 attack on Saudi Aramco that erased data on hard drives and replaced them with an image of a burning American flag. The attack unleashed sophisticated malware known as Shamoon and infected tens of thousands of computers, over three quarters, within the organization. The attack caused significant financial losses and damaged Saudi Arabia's reputation as a reliable energy provider. It was one of the first to emphasize the critical industry and raise concerns about its potential impact.

We have also seen several attacks on financial institutions. One notable example was in 2014 when JP Morgan Chase hackers compromised the personal information of millions of customers. Another significant incident happened in 2016 when a bank in Bangladesh was targeted, resulting in the theft of $81 million. Going back to 2012, we witnessed many large-scale denial-of-service attacks on financial institutions. Yahoo had a breach in August of 2013, and the impact was that three billion customer accounts were exposed.

There have been several attacks on social media companies. Two companies, Alibaba and Aadhar, tied to Alibaba, had over a billion Indian citizens' information compromised, including bank accounts. Then we've seen over the past Facebook with over 500 million accounts. LinkedIn experienced a very large data breach that included 700 million accounts being exposed. Marriott International had over 500 million customer accounts exposed. And then, even going back further than that, Myspace—yes, remember them—a widely used social media company in 2008, experienced significant breaches. Cyberattacks, as you can see, go back years.

More recently, some of the breaches that have occurred since I wrote my last book, *The Coming Cyber War,* have been significant, including the SolarWinds breach in late 2020 that was mentioned above and the Colonial Pipeline attack. Others include an attempt on a Florida city, Oldsmar's, water supply, in which the threat actors were attempting to poison the water supply. That was a breach that could have impacted numerous individuals in the state of Florida. There is speculation on how the attack began, whether it was an error with an employee or actually someone who infiltrated the city's systems.

We have witnessed government agencies from the Finnish Parliament to municipalities in Canada and public transportation in India. Even our U.S. House of Representatives has been impacted because the health care system they are part of was involved in a breach, and the online marketplace that administers health care plans for members of Congress and certain Capitol Hill staff members was affected. Those are some significant breaches, and there have been countless cyberattacks on countries that we've seen across the globe since late 2020.

To reiterate all the things that we've talked about—why you should care, why Americans should care, as well as corporations and boards—cyber incidents are increasing in number, scope, and impact resulting from an attack; the importance of cybersecurity and protecting our critical infrastructure as well as personal data cannot be overstated. The United States, as a global leader and technically advanced country, along with many of our allies, faces many multifaceted risks in cyberspace. This emphasis on why Americans should care is really a matter of national security and economic stability. Our collective understanding and addressing these risks is crucial to safeguarding our nation's interests now and in the future. Preserving our country's intellectual property and our status worldwide is a matter of national security.

The cyber realm has become a new battleground for states and non-state actors alike. Our dependence on network systems and critical infrastructure makes the United States very vulnerable to cyberattacks. State-sponsored attacks from hostile nations employ cyber capabilities to target the U.S. government, our military, and our intelligence agencies, and they seek to steal classified information, disrupt our critical operations, and compromise national defense strategies.

Protecting this data and ensuring the integrity of government systems is paramount to our national security. Critical infrastructure protection, including power grids, transportation systems, and financial networks, is increasingly interconnected with one another and reliant on digital technologies. Cyberattacks on these sectors of our critical infrastructure can have catastrophic consequences. Attacks like these are occurring almost daily; certainly, attempts are occurring numerous times a day. Successful attacks can lead to power outages, disrupt our transportation systems, and, as mentioned, our water supply systems. We've all felt the impact of the financial instability that these incidents can cause. Prioritizing cybersecurity measures will mitigate risk and strengthen our resilience.

As it relates to our nation's and allies' defense industries, each defense company plays a critical role in national security and supplying our militaries. Cyber threats target not only the government agencies themselves but also the defense contractors that are making our weapon systems. Across all aspects of military supply chains, there is a significant risk posed by disruption, destruction, and theft. By protecting our intellectual property and securing supply chains, we can continue to enhance our resilience against cyber espionage, ensuring the United States and its allies have U.S. military readiness and superiority.

The economic risks are even more drastic. In the globalized economy, cybersecurity is inseparable from economic stability. The interconnectedness of financial systems, intellectual property theft, and the rise of cybercrime pose immense challenges. More importantly, this aspect can affect us personally. Intellectual property drives economic growth and competitiveness. Cyber intrusions target American businesses and their trade secrets to research findings, both at a business and in our education system, our institutions, and all the proprietary information, causing billions of dollars in economic losses. Cyber espionage, state sponsors, cyber actors committing economic espionage, and even corporate rivals engage in economic espionage, seeking to gain a competitive advantage by stealing proprietary information and customer databases to commit fraud. Protecting those corporate interests is imperative for our economic stability.

Consumer trust can be eroded. When financial institutions or merchants have breaches with credit card data stolen, that's significant. When it involves a customer's identity, it is even worse. Sifting through what has happened to an individual, having to restore and clear your credit along with refunds of invalid transactions, and getting reissued documentation seems to be the norm. Consumers need to protect their personal information and lock up every aspect of their personal data so that attempts get alerted. Breaches that compromise consumer data erode our trust across the globe, stymie business growth, and can have a lasting impact on our economy.

Again, cybersecurity is no longer an optional concern but an imperative for the United States and our allies around the globe. By comprehending this as a national security and economic risk, a systemic risk across the board associated with cyber threats, we can understand the significance of cybersecurity in preserving our way of life and national interests. Investing in robust cybersecurity systems, tools, and measures, fostering that great relationship between public and private partnerships at the national and state levels, and then enhancing our research and development in the field, as well as training up our next generation of cybersecurity guardians, can protect our national security, protect our critical infrastructure, and ensure our economic stability.

It will take a concerted effort across the United States and our allies to effectively navigate the complex cybersecurity and cyberspace landscape and maintain our positions as global leaders in the digital age today and in the future. Cybersecurity has become a matter of protecting our way of life, national security interests, and global stability.

From the Desk of a CISO: A CISO's Perspective
The Next Generation Featuring Jake Crudgington

Jake Crudgington Bio: Jake is a first-year student at the University of Texas at San Antonio, Alvarez College of Business, double majoring in cybersecurity and management information systems. Jake aspires to learn about cybersecurity and technology with the hopes of developing products that companies need and improving their cybersecurity programs. Jake's passions are cybersecurity, product development, motorcycle riding, athletic training, snowboarding, thrifting, hanging out with friends, and traveling.

> *"Tell me and I forget. Teach me and I may remember.*
> *Involve me and I learn."*
> - Benjamin Franklin, inventor, statesman & Founding Father

"Never say never, because limits, like fears, are often just an illusion."
- Michael Jordan, NBA Hall of Famer, 6-time world champion, and entrepreneur

The Need to Invest in a Next Generation of Cybersecurity Leaders and the Cybersecurity Profession

In our broadening digital world, where technology permeates every aspect of our lives, from our everyday home lives to how we work, the importance of cybersecurity and people filling cybersecurity roles cannot be overstated.

In an article found in Infosecurity Magazine titled, "Cybersecurity Workforce Gap Grows by 26% in 2022," the article states, "The global cybersecurity workforce gap has increased by 26.2% compared to 2021, with 3.4 million more workers needed to secure assets effectively, according to the (I.S.C.)2 2022 Cybersecurity Workforce Study."

We are facing a national security and defense dilemma with greater needs in critical infrastructure, data privacy, and intellectual property infrastructure. There is a growing cyber threat landscape, with advanced technology becoming readily available to threat groups that want to cause the United States and our allies harm.

I asked Jake what caught his interest in the field of cybersecurity. You may be surprised to find out that it was not me. "I really like technology, computers, and similar technology, and especially the defense side of cybersecurity. In my 10th grade C.I.S. class, my teacher, Mr. Guidry, had us do an information security project. The project was to assess a store and figure out how it could be hacked, then come up with solutions, protection plans, and strategies to improve their cybersecurity posture." All across our country, we have high school, even younger, students touching on cybersecurity, but they may not even know that it is cybersecurity or what the field is about.

As stated in an article published by C.S.O. Online, Jeff Robbins, Practice Director, Security/Wireless, for Business Communications, Inc. (BCI), was quoted as saying, "The threat landscape has really changed in the past ten years. Today, we are up against some really efficient and technical cybercriminals. They're not just running scripts; they have sophisticated techniques with malicious payloads. Not everyone outside our industry understands the level of skill we need to counter in the world of cybersecurity and cybercrime. Cybercriminals are very motivated by the amount of money they can make. Well-known state-sponsored attacks are just one example."

When it comes to today's sophisticated attacks, times certainly have changed. We are continuously developing advanced technology, and that innovation will rapidly increase with AI (artificial intelligence). AI has the capability to outperform humans due to its lack of human emotions, such as fear of death or harm. Autonomous weapons are being developed for use in warfare. An AI-controlled weapon will not have any ramifications for dying; therefore, it will take risks that humans wouldn't. The lack of human emotion and the ability to sync with other common devices can create a swarm effect with devastating consequences in the wrong hands.

Jake responded to the question by saying how he felt about the difficulty of learning cybersecurity. He stated, "There are a vast amount of opportunities in cybersecurity, with many rabbit holes to fall into and a lot of niche job opportunities. The cybersecurity field has a lot of content to grasp and a ton of levels at which you need to understand the information. Encryption, for example, can be as basic as two-step authentication, a form of encryption, or it can be as complicated as understanding what ARMv8 is."

The United States White House and the Office of the National Cyber Director (ONCD) developed the National Cyber Workforce and Education Strategy (NCWES) to address national security and economic imperatives. "The National Cyber Workforce and Education Strategy is critical for the National Cybersecurity Strategy to be possible, as we must equip our people with the necessary skills to implement the cybersecurity vision laid out in the National Cybersecurity Strategy," said Acting Director of ONCD, Kemba Walden.

Further, by investing in education and training programs, the United States and our allies can bridge this gap and ensure that a competent workforce is ready to meet the challenges the future in a digital age and cyberspace presents. "Building and maintaining a strong cyber workforce cannot be achieved unless a cybersecurity career is within reach for any capable American who wishes to pursue it, and every organization with an unfilled position plays a part in training the next generation of cybersecurity talent," said United States President, Joe Biden.

Jake responded to a question related to his short-term and long-term aspirations in cybersecurity by stating, "I hope to gain my master's in cybersecurity and get an MBA, maybe a dual master's, have five to six different certs, gain expert knowledge about AI, cloud networking and security, and defensive strategies, as well as meet peers that can help guide me in my career. I want to intern at a top cyber company in the San Antonio, Houston, or Austin areas and maybe eventually work for one of the companies I interned at. In the peak part of my career, I want to be a CISO or a founding member of a company that is innovating cybersecurity products."

As we continue to innovate and technology advances throughout the world, the importance of cultivating a new generation of cybersecurity professionals cannot be overstated. Diversity, whether younger practitioners of different nationalities, backgrounds, or sexes, is particularly important as it helps to enhance problem-solving and creativity when dealing with ever-evolving threats. Encouraging all people groups to pursue careers in cybersecurity is important to bolster the United States and our allies' capabilities. These experts are the guardians of digital society, protecting critical infrastructure, data privacy, and national security. Their role is to defend against current threats and anticipate and adapt to the evolving landscape of cyber risks. By investing in education, training, and diversity, the United States and our allies can ensure we remain leaders in cybersecurity innovation and resilience.

When asked what he wants to know from other leaders in the industry, Jake responded, "What can set me apart right now and prepare me for summer internships? How do I figure out what career direction I should pursue in cyber? And what are the best groups or associations I should get involved in to become more knowledgeable and advance my career in cyber?"

CHAPTER 2

National Security and Systemic Risks

In an increasingly interconnected and digital world, the importance of cybersecurity cannot be overstated. The United States, as a global leader and technological powerhouse, faces multifaceted risks in cyberspace. I want to emphasize why Americans should deeply care about cybersecurity and focus on the implications for national security and economic stability. Understanding and addressing these risks are crucial to safeguarding the nation's interests and preserving its position in the global arena. In an era dominated by rapid technological advancements and an ever-expanding digital landscape, the importance of cybersecurity and safeguarding national security cannot be overstated.

As the United States faces an array of evolving threats ranging from state-sponsored cyber espionage to sophisticated attacks, it is imperative to recognize cybersecurity's vital role in ensuring the United States and our allies' safety and maintaining our strategic interests.

Exploring this in the multifaceted dimensions of cybersecurity and its critical role in America's national security is one of the things that is imperative to do.

When it comes to national security, cybersecurity is an integral part, as the cyber realm has become a new battleground for states and non-state actors alike. The dependence on network systems and critical infrastructure makes the United States and our allies vulnerable to cyber threats. Americans should care about cybersecurity from a national security perspective due to the changing nature of cyber warfare and the cyber threat landscape. The emergence of cyberspace as a new domain of warfare has fundamentally transformed the nature of conflicts. State-sponsored hackers, non-state actors, and cybercriminals exploit vulnerabilities in interconnected networks to disrupt critical infrastructure, compromise sensitive information, and wage cyber espionage campaigns.

The United States, as a global leader and technological powerhouse, is an attractive target for such threats. Our allies are also attractive targets.

Sophisticated state-sponsored attacks by hostile nations are driven by geopolitical motives and employ advanced cyber capabilities to target American government institutions, military infrastructure, and intelligence agencies. These attacks aim to steal sensitive information, compromise national defense strategies, and disrupt critical operations. America's cybersecurity posture serves as the first line of defense against such threats, ensuring the protection of classified data and preserving our national security interests.

Cyberterrorism and non-state actors are on the rise, and cyberterrorist organizations present a new dimension of threats to national security. These entities leverage cyberspace's anonymity and global reach to carry out attacks on critical infrastructure, disrupt public services, and undermine societal stability. A robust cybersecurity posture is crucial in detecting such attacks, minimizing the potential damage, and ensuring the safety of American citizens and our global allies. One of the unique characteristics of cyberspace is the inherent difficulty in attributing attacks to their perpetrators. Cyber adversaries can operate covertly, employing sophisticated techniques to mask their identities.

This challenges the traditional paradigms of deterrence, making it crucial for the United States to bolster its cybersecurity capabilities to identify, track, and respond to cyber threats. Protecting national security and safeguarding government systems, government networks, data, and communications is vital for national security. Cyber intrusions targeting government entities aim to compromise classified information, disrupt critical operations, and undermine the nation's ability to govern effectively. The United States can fortify its resilience against cyber threats by strengthening cybersecurity measures and adopting proactive defense strategies.

We've seen recently, in June 2023, that many U.S. government agencies have come under attack due to a single vulnerability, the MOVEit Transfer vulnerability. MOVEit Transfer is a managed file transfer software, and the vulnerability, formerly known as CVE 202-3578, is a privilege escalation vulnerability that a threat actor can exploit and take over systems that are impacted. Defending the nation's critical infrastructure, including energy grids, transportation systems, financial systems, public events, large structures, and telecommunications, relies heavily on digital technology, as do all critical infrastructure sectors. Cyberattacks targeting these sectors can have severe consequences, disrupting daily life, compromising public safety, and potentially causing widespread chaos. Robust cybersecurity measures are imperative to safeguard critical infrastructure from disruptive attacks and enhance the nation's ability to respond to emergencies. In protecting critical infrastructure, interconnected networks, and vulnerabilities, the United States relies heavily on interconnected networks, including energy grids, transportation systems, financial institutions, and healthcare facilities.

These sectors are potential targets for cyberattacks that could have devastating consequences for national security. Not only that, but many of these targets are also considered soft targets, as they have not taken cybersecurity seriously and have many vulnerabilities in their infrastructure. A strong cybersecurity posture is essential for protecting critical infrastructure, preventing disruptions to essential services, and maintaining societal functioning during times of crisis. Industrial control systems and cyber-physical threats are involved in the convergence of information technology and operational technology, which has introduced new cyber-physical threats. Attacks targeting industrial control systems, for short, can disrupt vital sectors such as energy, water, and manufacturing, with far-reaching consequences for national security.

America's cybersecurity posture again plays a crucial role in safeguarding these systems from threat intelligence in response to ensuring their integrity and preventing potentially catastrophic events. The nation's critical infrastructure, including power grids, transportation systems, and financial networks, is increasingly interconnected and reliant on digital technologies. Cyberattacks on these vital sectors can have dire consequences, leading to power outages, disrupted transportation, and financial instability. Prioritizing cybersecurity measures will mitigate such risks and strengthen the nation's resilience.

Next, I'll talk about military readiness and cyber warfare. Preserving military capabilities as technology increasingly becomes integrated into military operations and protecting defense networks, weapon systems, and military supply chains is paramount.

Cyberattacks targeting the defense sector seek to undermine military readiness, steal sensitive data, and compromise advanced weapon systems. Ensuring the integrity of military operations and safeguarding sensitive information, whether it is directly from the military or our contractors, is crucial to maintaining America's military superiority. The modern battlefield extends into cyberspace, and the United States must maintain a formidable cybersecurity posture to ensure military readiness. Protecting military networks, securing communication systems, and defending against cyberattacks on weapon systems are critical to preserving operational capabilities and maintaining an advantage over our adversaries. Another technique the military deploys is offensive cyber capability. The ability to conduct offensive cyber operations serves as a deterrent against potential adversaries. America's cybersecurity posture encompasses not only defensive measures but also the development of offensive cyber capabilities. These capabilities enable the United States and our allies to retaliate against cyberattacks, disrupt adversary networks, and impose consequences that safeguard national security interests.

Cybersecurity plays a critical role in deterring and retaliating against state-sponsored attacks. Developing robust offensive capabilities allows the United States to respond to attacks swiftly and proportionately, sending a clear message that hostile actions in cyberspace will not go unpunished. The ability to defend national interests in cyberspace bolsters overall national security posture and ensures a credible deterrent against potential adversaries.

Our defense industry plays a crucial role in national security. Cyber threats targeting defense contractors, weapon systems, and military supply chains pose significant risks. The F22 Raptor had well over 1000 supplier firms, and that may just be scratching the surface when you include subcontractors. By protecting this intellectual property, securing supply chains, and enhancing resilience against cyber espionage, the U.S. can ensure its military readiness.

To counter state-sponsored threats, we have to look at state-sponsored cyber espionage. Those are hostile nations seeking to gain strategic advantages and conduct cyber espionage campaigns against the United States by infiltrating government networks, defense contractors, and/or research institutions. These adversaries aim to steal classified information, our intellectual property, and the technological innovations that we deploy. Effective cybersecurity measures are essential to protect national secrets, preserve economic competitiveness, and counter the threats posed by rival nations.

State-sponsored attacks employ cyber capabilities to target the U.S. government, military, and intelligence agencies. They seek to steal the information that these entities hold, disrupt critical operations, and compromise national defense strategies.

Protecting this sensitive data and ensuring the integrity of government systems are paramount to safeguarding national security.

When it comes to intelligence and counterintelligence operations, securing classified information is paramount. Government agencies and military organizations rely on secure communication channels, encrypted databases, and protected networks to safeguard classified information. The cybersecurity posture of the United States directly impacts its ability to protect intelligence assets, prevent espionage, and maintain the confidentiality of sensitive data.

A robust cybersecurity framework is critical to preventing unauthorized access, detecting insider threats, and preserving national security secrets. Cyber espionage and foreign influence are persistent threats to national security, aiming to steal intellectual property, compromise defense capabilities, and influence the political process.

By continuously strengthening our cybersecurity posture, the United States and our allies can enhance our ability to detect and respond to cyber espionage, protect critical research and development, and counter foreign influence campaigns. This highlights the need for collaboration and innovation. Public-private partnerships are imperative to accomplish this. Collaboration between the public and private sectors is crucial to addressing cybersecurity challenges. Close cooperation between government agencies such as the FBI, through its InfraGard program, industry leaders, and technology providers, enables sharing of threat intelligence, best practices, and resources. Public-private partnerships foster innovation, enhance information sharing, and collectively strengthen the nation's cybersecurity defense.

When we invest in research and development to maintain a competitive edge and effectively counter emerging cyber threats, sustained investment in these programs is vital to our national interest. Advancements in artificial intelligence, machine learning, and quantum computing, which we'll discuss in later chapters, hold promise for bolstering the nation's cyber capabilities. By fostering innovation and supporting academic institutions, the United States can cultivate a skilled workforce and stay at the forefront of cybersecurity advancements.

Investing in the future cyber-skilled workforce through university programs, high school programs, and many other programs is paramount to our continued strengthening of that posture. Now, let's talk about economic risk. In today's globalized economy, cybersecurity is inseparable from economic stability. The interconnectedness of global economies has brought forth immense opportunities for economic growth and innovation. It has also crippled financial institutions, resulted in numerous accounts of intellectual property theft, and seen a rise in cybercrime. Nations, including the United States and many of our allies, are exposed to unprecedented economic risks stemming from cybersecurity threats. There are financial losses, intellectual property theft, disruption of critical industries, erosion of consumer trust, and the cost of cybersecurity measures, all coming into play when it comes to economic risk.

When it comes to intellectual property protection, innovation, and intellectual property drive our economic growth and competitiveness. Cyber intrusions targeting American businesses seek to steal trade secrets, research findings, and proprietary information, causing billions of dollars in economic losses. Ensuring robust cybersecurity measures protects the nation's economic vitality and supports job creation.

According to the 2023 World Economic Forums, otherwise known as the W.E.F. Global Risk Report, cybersecurity is a top-ten global risk in the near- and long-term future. It is also expected that by 2025, cybercrime will have an annual hit or annual cost of $10.5 trillion.

State-sponsored cyber actors and corporate rivals engage in economic espionage, seeking to gain a competitive advantage by stealing proprietary information, customer databases, and/or research and development data. Protecting these interests and trade secrets is essential to preserving economic sovereignty. The theft of intellectual property, often conducted by state-sponsored entities, targets American corporations and research institutions to steal that intellectual property and associated trade secrets.

This theft of valuable innovations and proprietary information undermines our competitive advantage as well as that of our allies. It also stifles innovation and hampers our economic growth. The loss of intellectual property diminishes incentives for research and development, ultimately impacting the nation's ability to remain at the forefront of technological advancements. Another ramification is the counterfeit goods and piracy that take place. Cyberattacks can facilitate the proliferation of counterfeit goods and digital piracy, negatively impacting industries such as pharmaceuticals, luxury goods, entertainment, and software. Counterfeit products not only harm legitimate businesses but also endanger consumer health and safety. This undermines market confidence, erodes brand value, and leads to revenue losses for legitimate industries, affecting employment and overall economic performance.

Another area of concern is the disruption of these critical industries. Critical infrastructure sectors, including energy, transportation, and healthcare, are increasingly reliant on digital systems. Cyberattacks targeting these sectors can disrupt operations, cause service outages, lead to significant economic consequences, and cause harm to citizens. For instance, attacks on energy grids can disrupt power supply, impacting businesses, consumers, and essential services. The resulting economic disruptions can have cascading effects on multiple sectors of the economy. Cyberattacks can disrupt business operations, leading to significant economic losses. Downtime resulting from system outages, ransomware attacks, or data breaches can halt production, disrupt supply chains, impact revenue generation, and have more drastic effects, such as harming our citizens.

As we've seen in the attack on Florida's water supply, small and medium-sized businesses, in particular, may struggle to recover from financial blows, potentially leading to layoffs, closures, and broader economic repercussions. Another ramification that disrupts our economy is cyberattacks on supply chains that can disrupt the flow of goods and services, impacting businesses across various industries.

Compromised supply chains can result in delays, increased costs, and product shortages. The economic repercussions extend beyond the immediate impact, affecting downstream suppliers, retailers, and consumers. Such can lead to reduced consumer spending, reduced business investments, and overall economic instability. Another impact is the erosion of consumer trust and confidence. As online transactions and digital services become the norm, consumers need to trust that their personal information, financial data, and online identities are secured. Cyber breaches that compromise consumer data erode trust, undermine business growth, and have lasting economic implications. Prioritizing cybersecurity enhances consumer confidence and fosters a thriving digital economy. High-profile data breaches undermine consumer trust in businesses and online services.

The exposure of personal information erodes confidence in the digital economy, leading to a reluctance to engage in online transactions and share sensitive data. This loss of consumer trust in the U.S. and our allied nations can result in reduced e-commerce activity, hamper business growth, and impede the digital transformation of industries. There is also an impact on e-commerce. The growth of e-commerce is pivotal to the American economy. Cybersecurity risks such as payment fraud, identity theft, and phishing attacks can hinder consumer adoption of online shopping platforms. A lack of confidence in the security of online transactions restricts market expansion, limits revenue potential for businesses, and inhibits job creation in the digital economy.

There is also a cost associated with implementing cybersecurity measures. The increased spending on cybersecurity is necessary to mitigate cybersecurity risk. Businesses and government entities are compelled to invest in robust cybersecurity measures, including advanced technologies, employee training, and incident response capabilities. The cost of implementing these measures can strain budgets, particularly for small businesses and government agencies with limited resources.

Increased spending on cybersecurity diverts funds that could otherwise be allocated to innovation, research, and economic growth initiatives.

Though each of these is imperative to the cybersecurity attacks we are constantly under, another impact related to cybersecurity costs is the opportunity cost we lose by focusing on cybersecurity. We divert attention and resources from core business activities, limiting productivity and innovation.

The opportunity costs associated with prioritizing defensive measures Over strategic investments can hinder long-term economic competitiveness and technological advances. Though there is opportunity cost, it is necessary to continue to bolster our cyber defenses to thwart attacks, preserve our economic way of life, and preserve the intellectual property that these investments can aid. Safeguarding personal data is also important when it comes to consumer trust and confidence.

Cybersecurity is essential for safeguarding personal data and the e-commerce systems we have come to rely on. As cybercriminals continue to steal personal and financial information, including Social Security numbers, credit card details, and medical records, trust in those systems erodes and puts an undue burden on entities involved in our financial systems. Robust cybersecurity practices that protect personal data are critical for ensuring the privacy and security of individuals and preserving trust in our systems. They are also vital in enabling the business to innovate and prosper. Cybersecurity is also important for our international relations.

Cybersecurity plays a pivotal role in shaping and safeguarding international relations. Cyberattacks can have diplomatic consequences. Cyber espionage, cyber terrorism, and state-sponsored attacks can lead to tension and conflicts between nations, making cybersecurity a national and international concern.

America's and other global allies' strong cybersecurity stance is important for building trust among nations, protecting against cyber threats from nations that threaten our way of life, preserving sovereignty, promoting diplomatic cooperation, and maintaining global stability. Cybersecurity is essential for fostering trust among nations in cyberspace. By demonstrating strong cybersecurity measures, countries can instill confidence in their counterparts, ensuring the protection of shared information, sensitive data, and critical infrastructure. Trust building in cyberspace is crucial to establishing stable and cooperative relationships essential for international cooperation and diplomatic endeavors.

When it comes to protecting confidentiality and privacy, robust cybersecurity measures protect the diplomatic communications and exchanges our nation undertakes. Nations must have confidence that their confidential information and diplomatic discussions remain secure from unauthorized access and cyber espionage. Such protection fosters an environment of openness, transparency, and trust in international relations. State-sponsored cyberattacks pose significant challenges to international relations.

By investing in cybersecurity capabilities, countries can defend against state-sponsored cyberattacks and protect their critical infrastructure, national defense systems, and sensitive information. Strong cybersecurity measures act as a deterrent, sending a clear message that cyber aggression will be met with appropriate defenses across our allies and United States capabilities. Transnational cybercriminal activities, including hacking, financial fraud, and data breaches, transcend national borders and impact multiple countries. Cooperation among nations is essential for sharing threat intelligence, coordinating cyber responses, and apprehending cybercriminals. By working together, countries can combat cybercrime effectively, protect their citizens, and preserve the integrity of global digital networks.

Another measure for protecting international relations is related to sovereignty and national security. Preserving sovereignty in cyberspace is integral to preserving relationships in the digital realm. Just as countries protect their physical borders, they must safeguard their cyber borders from unauthorized intrusions, data theft, and cyberattacks. Each nation has the right to defend its digital infrastructure and regulate cyberspace within its jurisdiction, ensuring the preservation of its national interest in sovereignty. Cybersecurity is crucial for safeguarding national security interests. Cyberattacks can disrupt critical infrastructure, compromise defense capabilities, and undermine the integrity of military systems. By prioritizing cybersecurity, nations can protect their national security assets, prevent unauthorized access to sensitive information, and maintain strategic advantage in an evolving threat landscape.

Promoting diplomatic cooperation is also a vital component related to international relations. International cooperation on cybersecurity norms and standards promotes diplomatic collaboration, establishes agreed-upon norms and principles of behavior in cyberspace, fosters predictability and stability, and requires responsible states to conduct collaborative efforts in developing and adhering to these norms, enhance mutual understanding, facilitate dialogue, and pave the way for diplomatic resolutions to cyber-related disputes. Cybersecurity capacity-building initiatives promote international cooperation by assisting less developed nations in strengthening their cyber defenses. By providing technical assistance, training programs, and knowledge sharing, more advanced nations can help build the cybersecurity capabilities of less-resourced countries. Such collaboration efforts promote inclusivity, bridge the digital void, and ensure the stability of the global digital ecosystem.

The maintenance of global stability is also essential. Robust cybersecurity measures contribute to the prevention of cyber-related, cyber-enabled conflicts between nations. By enhancing defensive capabilities, establishing communication channels, and fostering transparency, countries can mitigate the risk of miscommunication, miscalculation, and unintended escalation in cyberspace.

The maintenance of global stability is dependent on a collective commitment to cybersecurity as a means to prevent conflicts. It is also crucial to mitigate economic disruptions that may arise from cyberattacks. A major cyber incident affecting one nation can have ripple effects across interconnected economies, impacting trade, supply chains, and financial systems.

By collaborating across nations on cybersecurity best practices, the United States and our allies can collectively reduce the economic vulnerabilities associated with cyber threats. Ensuring global economic stability through cybersecurity is vital for international relations, serving as a foundation for trust building, defense against cyber threats, preservation of sovereignty, promotion of diplomatic cooperation, and maintenance of global stability.

Given the interconnected nature of digital networks, nations must prioritize cybersecurity as a core element of their foreign policy agenda. Through collaborative efforts and adherence to cybersecurity standards, nations can promote a more secure and stable international digital landscape. This facilitates diplomatic cooperation and safeguards shared interests in the digital age.

Another impact is regulatory mandates. In an era of increased cyber threats and the critical importance of safeguarding digital assets, the United States and many of our global allies have developed comprehensive regulatory frameworks to address cybersecurity and data privacy concerns. Regulations are important to protect critical infrastructure, ensure data privacy, promote industry best practices, and foster a resilient and secure cyber ecosystem. However, it is important to note that no regulation should be the end-all for cybersecurity defenses. Each organization must measure its risks and develop cyber practices to mitigate them.

Focusing on some of the more recent regulations, we have seen:

1. *The Cybersecurity Enhancement Act of 2014.* This act focuses on strengthening cybersecurity research and development, promoting public-private partnerships, and advancing the sharing of cyber threat information. It encourages collaboration between government agencies, industry stakeholders, and academia to enhance the nation's cybersecurity capabilities.

2. *The Federal Information Security Modernization Act,* known as FISMA, was created in 2014. FISMA mandates federal agencies to develop, implement, and maintain information security programs. It requires agencies to conduct regular assessments, establish incident response capabilities, and report on their cybersecurity posture. FISMA emphasizes continuous monitoring, vulnerability management, and the protection of sensitive government information.

3. *The Cybersecurity Information Sharing Act, CISA, of 2015* promotes the sharing of cyber threat information between government and private sector entities. It enables the exchange of

actionable intelligence, facilitates timely incident response, and enhances the overall cyber defense posture of the nation. CISA encourages collaboration while safeguarding privacy rights and ensuring the protection of shared information.

There are also regulatory standards. The National Institute of Standards and Technology Cybersecurity Framework provides a voluntary set of standards, guidelines, and best practices for organizations to manage and mitigate cybersecurity risk. It offers a flexible approach to aligning cybersecurity efforts with business objectives, fostering risk-based decision-making, and promoting a proactive security culture. The framework is widely adopted by both public and private sector organizations.

4. An additional act relating to our health care, known as *The Health Insurance Portability and Accountability Act,* or HIPAA for short, establishes standards for the protection of sensitive patient health information within the healthcare industry. It requires covered entities to implement safeguards to ensure the confidentiality, integrity, and availability of protected health information. HIPAA regulations aim to protect individuals' privacy rights while facilitating the secure exchange of healthcare data across healthcare companies.

5. Another standard to protect individual privacy is *The Payment Card Industry Data Security Standard.* Otherwise known as PCI DSS. The PCI DSS standard is a set of requirements designed to ensure the secure handling of credit card information. It applies to organizations that process, store, or transmit credit cardholder data. Compliance with PCI DSS helps prevent payment card fraud and protects the confidentiality of cardholder data.

Regulatory agencies that help protect our data include:

1. *The Department of Homeland Security.* The Department of Homeland Security (DHS), plays a crucial role in the regulation and coordination of cybersecurity efforts across various sectors. DHS oversees critical infrastructure protection, leads national security response efforts, and performs assessments for the industries they protect. It also provides guidance on best practices related to cybersecurity and the protection of critical infrastructure. DHS collaborates with other federal agencies, private sector partners, and international counterparts to address emerging cyber threats.

2. *The Securities and Exchange Commission* (SEC) is also proposing rules to enhance and standardize disclosure regarding cybersecurity risk management, strategy, governance, and cybersecurity incident reporting by public companies that are subject to the reporting requirements of the Security Exchange Act of 1934. Specifically, the SEC is proposing amendments to require current reporting about material cybersecurity incidents. The SEC is also proposing to require periodic disclosures about registrants, policies, and procedures to identify and manage cybersecurity risk management's role in implementing cybersecurity policies and procedures and the board of directors' cybersecurity expertise, if any, and its oversight of cybersecurity risk.

3. *The Federal Trade Commission* (FTC) has the authority to enforce regulations related to consumer privacy, consumer data privacy, and cybersecurity. The FTC acts against organizations that engage in unfair or deceptive practices concerning the protection

of consumer information. The FTC promotes transparency, data breach notification, and businesses' adoption of reasonable security measures.

4. Another government agency is the *Federal Communications Commission*. The FCC focuses on cybersecurity regulations related to the telecommunications industry. It ensures the protection of telecommunications infrastructure, promotes secure networks and addresses cybersecurity issues in the realm of broadband services and wireless communication.

There are also global regulations to consider. While there is no comprehensive global cybersecurity regulation framework, several international initiatives and agreements contribute to the establishment of common cybersecurity principles and guidelines. Some notable global cybersecurity regulations include:

1. *United Nations General Assembly Resolutions*: The United Nations, UN, has adopted various resolutions highlighting the importance of cybersecurity and encouraging member states to collaborate in addressing cyber threats. Resolutions such as UN General Assembly resolutions 70/237 and 73/266 emphasize the promotion of a secure and trustworthy cyberspace, respect for human rights online, and the need for international cooperation in combating cybercrime.

2. *The International Telecommunication Union* (ITU) is a specialized agency of the United Nations. They play a significant role in developing global cybersecurity regulations. The ITU focuses on enhancing the security and resilience of information and communications technologies and promoting international collaboration. The ITU's Global Cybersecurity Agenda provides

a framework for addressing cybersecurity challenges and facilitates the exchange of best practices.

3. *The Council of Europe Convention on Cybercrime.* The Council of Europe Convention on Cybercrime, also known as the Budapest Convention, is an international treaty that aims to harmonize cybercrime legislation and international cooperation in investigating and prosecuting cybercriminal activities. It addresses a wide range of cyber offenses, including illegal access, data interference, and computer-related fraud, and encourages member states to establish effective legal frameworks to combat cybercrime.

4. *The European Union General Data Protection Regulation (GDPR).* While not specifically focused on cybersecurity, the GDPR sets strict rules regarding the protection of personal data and the rights of individuals. It applies to all EU Member States and has extraterritorial reach, impacting organizations worldwide that process the personal data of EU citizens. The GDPR emphasizes the need for strong cybersecurity measures, data breach notification, and the protection of individual rights.

5. *The Asia-Pacific Economic Cooperation Privacy Framework.* The APEC Privacy Framework provides guidance for member economies in promoting privacy protection in the digital environment. The APEC Privacy Framework emphasizes the importance of personal data protection, cross-border data flows, and cooperation among member economies to address privacy-related challenges. The framework aims to foster trust and promote responsible data-handling practices in the Asia-Pacific region.

6. *The International Organization for Standardization* (ISO). The ISO develops international standards that address various aspects of cybersecurity. The ISO/IEC 2700 Series, particularly the 27001 Information Security Management System and the 27002 Code of Practice for Information Security Controls, provide guidance for organizations in establishing and maintaining effective information security systems. These standards are widely adopted and recognized globally.

Many joint collaboration activities exist between governments, private sector organizations, and international bodies contributing to global cybersecurity regulations.

One example is the World Economic Forum's Partnering for Cyber Resilience Platform, which fosters public-private cooperation in addressing cybersecurity challenges. The Global Forum on Cyber Expertise facilitates knowledge sharing and capacity building among countries to strengthen cybersecurity capabilities worldwide.

The United States and many of our allies have established robust regulatory frameworks to address cybersecurity challenges and protect critical public and private assets. These regulations, encompassing legislative acts, standards, and regulatory agencies, aim to foster a secure and resilient cyber ecosystem. By adhering to these regulations, organizations can mitigate cyber risk, safeguard sensitive data, and contribute to the overall cyber defense of the United States and our allied nations. Continuous adoption and collaboration between stakeholders will be crucial in effectively addressing emerging cyber threats and ensuring the effectiveness of American and global cybersecurity regulations in the face of an evolving threat landscape. In conclusion, I'd like to drive home the sense of urgency we, the United States of America, and our global allies are facing.

A rallying cry to raise the alarm and the awareness level that all our critical infrastructure is under attack, plain and simple. Cybersecurity is no longer optional but imperative for the United States and our allies.

By comprehending the national security and economic risks associated with cyber threats, Americans and citizens across the globe can understand the significance of cybersecurity in preserving our way of life. Investing in robust cybersecurity measures, fostering public-private partnerships, and advancing research and development are essential steps to safeguarding national security, protecting critical infrastructure, and ensuring economic stability. Only through conservative efforts can the United States effectively navigate the complex cyberspace landscape and maintain its position as a global leader.

Cybersecurity risks pose significant economic challenges to the United States and our global allies in the digital age. These are systemic risks we must take seriously and overcome. Financial losses, intellectual property theft, disruption of critical industries, erosion of consumer trust, and the cost of cybersecurity measures all contribute to the economic risk faced by the nation. It is imperative for businesses, government entities, and individuals to recognize the magnitude of these risks and invest in proactive cybersecurity measures to mitigate the economic impacts, as well as design very strong recovery strategies. Strengthening cybersecurity resilience is essential for protecting the nation's economy and our allies, fostering innovation, ensuring market competitiveness, and sustaining long-term economic growth.

Recognizing the ever-evolving threat landscape and the transformative nature of cyberspace, the United States and our allies must prioritize the protection of government systems, critical infrastructure, and military capabilities. By countering state-sponsored threats, fostering collaboration, and investing in research and development, we can fortify our cybersecurity defenses and ensure our citizens' safety, sovereignty, and prosperity. A comprehensive and proactive approach to cybersecurity is important to maintain our position as a global leader in an increasingly interconnected and digital world and to help protect our allies and preserve democracy around the globe.

CHAPTER 3

Cyber Threat Actor Landscape and Motives

In this chapter, I'd like to discuss the various types of threat actors. These are some of the most sinister actors across the globe. While this is a little in the weeds, so to speak, I do believe the better individuals and organizations understand the adversaries and threats that we all face, the better we can prepare. In this age of data sprawl devices proliferating to the edge of infrastructure and personal information related to just about everything spread across infrastructures, cyber threats have risen to one of the most significant risks that individuals, organizations, and governments encounter, both in the United States and across the globe.

Behind these threats are diverse types of cyber threat actors, each with distinct motivations, capabilities, and techniques. Understanding these actors is crucial for developing effective cybersecurity strategies and mitigating the risks they pose. This chapter delves into the different types of cyber threat actors, including state-sponsored hackers, criminal cybercriminals, hacktivists, and insider threats, along with other types of threats.

A report published by Palo Alto Networks titled 2022 Unit 42 Network Threat Trends Research Report Volume One noted that the ratio of malware samples to benign files nearly doubled from the previous twelve months. The rapid pace of threats, malware exploits, and attacks is increasing at an alarming rate. As a service type, attacks are readily available on the Dark Web. From ransomware as a service to reconnaissance, anyone with a motive and a little cryptocurrency can carry out an attack.

State-sponsored hackers and certain criminal gangs are backed by nation-states and carry out cyberattacks to achieve political, economic, or military objectives. Their motivations can include espionage, intellectual property theft, disruption of critical infrastructure, and gaining a strategic advantage over rival nations. State-sponsored hackers often possess advanced technical capabilities and significant resources. They employ sophisticated techniques such as zero-day exploits, advanced persistent threats, and targeted attacks. They may also engage in covert operations, including disinformation campaigns and influence operations.

Notable examples of state-sponsored cyber threat actors include:

- APT (Advanced Persistent Threat) 29, otherwise known as Cozy Bear.
- APT 28, known as Fancy Bear, is associated with Russia.
- The Lazarus group is linked to North Korea.
- The Equation group is believed to be associated with the United States.

Here are other state-sponsored threat actors:

First is AlphaV, aka BlackCat, aka Alpha Spider, who is the creator of BlackCat Ransomware. They are a relatively new and rapidly growing cybercrime group. First observed near the end of 2021, the AlphaV Group gained attention for innovative extortion tactics and unconventional attack methods. They are widely known as the operators of the distinctly nefarious ransomware as a service offering known as BlackCat.

On April 19, 2022, the Federal Bureau of Investigations published a flash alert about the BlackCat AlphaV activities on March 22.

According to the FBI's report, BlackCat is believed to be the successor of Revel, DarkSide, and BlackMatter ransomware operators. The FBI further stated in the FLASH report, "BlackCat/ALPHV ransomware leverages previously compromised user credentials to gain initial access to the victim system. Once the malware establishes access, it compromises Active Directory user and administrator accounts." An article that appeared in a widely read cybersecurity online site stated, "Discovered in November 2021, the group was feared for its sophistication. Experts and researchers believe the group may be associated with other advanced-persistent threat (APT) groups like Conti, DarkSide, REvil, and BlackMatter."

In May 2021, Colonial Pipeline was targeted by a ransomware attack, which resulted in the shutdown of the entire pipeline system responsible for distributing gasoline and jet fuel to the East Coast of the United States. This caused fuel shortages and price increases in several states along the area. Colonial Pipeline had to pay nearly $5 million for the ransom, and after the investigation, the FBI confirmed that DarkSide ransomware was responsible for the compromise. AlphaV has two distinct competitive advantages over other ransomware-as-a-service subscription operators. First, BlackCat is the first widely distributed ransomware family to be written in Rust, a programming language that allows the operator to easily customize malware against different operating systems.

One of the benefits threat actors take advantage of when using Rust is that it adds another level of difficulty during analysis due to its obscurity. Obscurity creates stealth. The sheer number of enterprise environments open to attack is an attractive quality for affiliates. The AlphaV threat group uses double and triple extortion tactics. This indicates that in addition to encrypting data and systems and threatening to leak exfiltrated data, AlphaV sometimes additionally threatens to launch distributed denial of service attacks, otherwise known as DDoS attacks, against the websites of victims who refuse to pay, thus completing the extortion trifecta.

The Alpha Group does not appear to target a specific sector or country. Because AlphaV allows other threat actors to use its Black Cat ransomware on a subscription basis, the presence of the malware on a system does not necessarily indicate a direct attack by AlphaV. To date, BlackCat ransomware has struck retail, financial, manufacturing, government, technology, education, and transportation industries across various countries, including the United States, Australia, Japan, Italy, Indonesia, India, and Germany.

The next group is APT38, otherwise known as Lazarus, Hidden Cobra, and Guardians of Peace, who are a team, and Zinc. This group is associated with North Korea and is known for perhaps the biggest cyber heist of all time, the attack on the Bangladesh Bank, which led to the theft of more than $81 million in February 2016. Yet the group has done much more than that. Mandiant, a widely recognized cybersecurity firm that is now part of Google Cloud, stated, "This large and prolific group uses a variety of custom malware families, including backdoors, tunnelers, data miners, and destructive malware, to steal millions of dollars from financial institutions and render victim networks inoperable." Lazarus has been behind numerous operations in the past decade, starting with DDoS attacks against South Korean websites, then moving on to targeting financial organizations and infrastructure in South Korea, continuing with the attack on Sony Pictures in 2014 and the launch of WannaCry ransomware in 2017.

In recent years, Lazarus started looking into ransomware and cryptocurrency, and it also targeted security researchers to gain information about ongoing vulnerability research. This group has unlimited resources and very good social engineering skills.

These social engineering skills were put to work during the ongoing COVID-19 health crisis, when pharmaceutical companies, including vaccine makers, became some of Lazarus's most urgent targets. According to Microsoft, the hackers sent spear-phishing emails that included fabricated job descriptions, luring their targets into clicking on malicious links. This group differs from others because, while it is state-sponsored, its targets are not governments but businesses and sometimes individuals who may have information or access that North Korean spies might want to get their hands on. Lazarus uses a variety of custom malware families, including backdoor tunnelers, data miners, and destructive malware, sometimes developed in-house. They spare no effort in relentless campaigns.

APT38 is unique in that they are not afraid to aggressively destroy evidence or victim networks as part of its operations. This group is careful and calculated and has demonstrated a desire to maintain access to victim environments for as long as necessary to understand the network layout, pivot from location to location, gain required permissions, and gain system technologies to achieve their goals.

The next threat actor is UNC2452, merged into APT29, also known as Dark Halo, Nobelium, SilverFish, and StellarParticle, and is a Russia-based espionage group assessed to be sponsored by the Russian Foreign Intelligence Service (SVR). They are most notably attributed to the SolarWinds breach. Mandiant stated, "The group name used to track the SolarWinds compromise in December 2020 is attributable to APT29." In 2020, thousands of organizations downloaded a tangled software update of the SolarWinds Orion software, giving the actor a point of entry into their systems. The Pentagon, the UK Government, the European Parliament, and several government agencies and companies across the world fell victim to this supply chain attack. The cyber espionage operation had gone unnoticed for at least nine months before it was discovered on December 8, 2020, when the security company FireEye announced it was a victim of a state-sponsored attacker that stole several of its Red Team tools. This hack proved more extensive than initially thought. The supply chain attack on the SolarWinds Orion software was just one entrance channel used by the attacker. Researchers found another supply chain attack, this time on Microsoft Cloud services.

They also noticed that several flaws in Microsoft and VMware products were exploited. UNC2452 is one of the most advanced, disciplined, elusive, and dangerous threat actors that is currently being tracked. Their tradecraft is exceptional. They have mastered both offensive and defensive skills and have used these techniques and tactics to refine their intrusion techniques to hide right in plain sight. They demonstrate a level of operational security that is rarely seen by other threat actors and spend much time inside government agencies and companies without being noticed. The NSA, the National Security Agency, the FBI, the Federal Bureau of Investigations, and a few other U.S. agencies said that the operation of their attack was sponsored by Russia against the US. Then, they imposed sanctions against Russia. There are also clues that point to the Cozy Bear APT29 Group.

The next threat actor group to mention is APT33. This group is suspected to be from Iran. They generally target aerospace and energy. They have targeted organizations spanning multiple industries, including those headquartered in the United States, Saudi Arabia, South Korea, and many others. APT33 has shown particular interest in organizations in the aviation sector involved in both military and commercial capacity and organizations in the energy sector with ties to petrochemical production. There are certain types of malware that are associated with this particular threat actor group. ShapeShift, DropShot, TurnedUp, NanoCore NetWire, and AlphaShare are all some of the malware strains that they may use.

"APT33 sends spear phishing emails to employees whose jobs are related to the aviation industry. These emails are recruitment-themed, contain lures, and contain various links to malicious HTML application files. The files contain job descriptions and links to legitimate job postings on popular employment websites that would be relevant to the target individuals." This is one of the attack vectors this threat group uses.

Another critically dangerous threat group is APT37, which is associated with North Korea. They primarily target South Korea, though they also target Japan, Vietnam, and the Middle East in various industry verticals, including chemical, electronics, manufacturing, aerospace, automotive, and healthcare. "APT37's recent activity reveals that the group's operations are expanding in scope and sophistication with a toolset that includes access to zero-day vulnerabilities and a specific malware known as wiper malware." Numerous cybersecurity bodies and reports claim the activity of APT37 is carried out on behalf of the North Korean government, given malware development artifacts and targeting that align with North Korean state interest and the dire need for North Korea to generate cash flow.

The associated malware with this group is a diverse set of malware for initial intrusion and exfiltration. They also develop custom malware that is used for espionage purposes, and they have access to destructive malware.

The attack vectors that APT37 uses include social engineering tactics tailored to specifically desired targets, strategic web compromises typical of targeted espionage, cyber operations, and the use of file-sharing sites to distribute malware more indiscriminately. They frequently exploit vulnerabilities in the Hangul word processor and Adobe Flash, according to Mandiant.

Another threat actor group is APT32, also known as the OceanLotus threat group. This threat group is believed to be based in Vietnam and has conducted malicious cyber operations since at least 2014. "This threat actor, known to use watering-hole attacks to compromise victims, targets organizations of interest to the Vietnamese government for espionage purposes." Its targets have included various private industries, foreign governments, and individuals such as dissidents and journalists, with a particular focus on entities operating in Southeast Asian nations, including Vietnam, the Philippines, Laos, Cambodia, and Thailand.

APT32 frequently employs tactics such as strategic web compromise to gain access to victim systems. Among other industries that this group has attacked are defense and technology groups, many manufacturing industry companies, and healthcare companies. This team is known for using a suite of remote access trojans (RATs) to leverage new network attack capabilities. The group also uses steganography, the technique of hiding secret data within an ordinary, non-secret file or message, to embed payloads in images.

Other attack vectors they use are to leverage active mine files that employ social engineering methods to entice the victim into enabling macros. Upon execution, the initialized file downloads multiple malicious payloads from a remote server. The threat actors then deliver malicious attachments via spear phishing emails. Evidence has shown that some are sent across various widely used email programs. The associated malware with this group are SoundBite, Windshield, Phoreal, Beacon, and Komprogo.

Another threat actor group is Mustang Panda. This group made numerous headlines in late 2022. This threat actor group is known as being based in China, and they leverage legitimate applications to target the Southeast Asian state of Mongolia. This threat actor group, Mustang Panda, has also leveraged the Russian-Ukrainian war to attack targets in Europe and the Asia-Pacific region. The group was first detected in 2017. Many believe they've been active since 2014. They have since targeted a wide range of organizations, including government agencies, nonprofits, religious institutions, and non-government organizations around the globe, in countries such as the United States, the European Union, Pakistan, and Vietnam. They frequently use tactics such as PlugX and China Chopper for their operations. "The research also highlights the "alarming" role USB drives play in spreading malware quickly and often unbeknownst to users—even across air-gapped systems," as noted in a DarkReading article.

PlugX is a remote access trojan that can be configured to use both HTTP and DNS for command-and-control activities. China Chopper is a malicious web shell that allows unauthorized access to an organization's network.

Another threat actor group is APT29, also known as The Dudes, which is merging with UNC2452. This group's motive is data theft. "They typically use custom malware families, including SUNBURST, BEACON droppers, RAINDROP, and TEARDROP, a credential theft tool called MAMADOGS, and CRIMSONBOX, a .NET tool that extracts the token signing certificate from an ADFS configuration, assisting the group in forging SAML tokens," according to Mandiant.

APT25 has historically used spear phishing as its attack vector, including messages containing malicious attachments and links embedded in its emails. They typically do not use zero-day exploits but may leverage those exploits once they have been made public. Associated malware with APT25 include Lingbo, PlayWork, MadWolf, and others.

APT24, also known as PittyTiger, is a threat group suspected to be based out of China, and they've targeted a wide variety of industries, including organizations in the government, healthcare, construction, engineering, nonprofit, and telecom industries. The group is known to have also targeted organizations headquartered in the US and Taiwan. Historically, they used a utility to encrypt and compress stolen data to transfer it out of networks. Data theft associated with this threat actor mainly focuses on documents with political significance, suggesting they monitor the position of various nation-states on issues related to China's ongoing territorial or sovereignty disputes.

They typically use malware known as PittyTiger, Enfal, and Taidoor. Primarily, they use phishing emails, renewable energy, or business strategy themes as lures in those phishing emails. They often engage in cyber opera operations where the goal is intellectual property theft, usually focusing on data and projects that make a particular organization competitive in the industry in which they operate.

Another actor group is TA505, often associated with FIN11, a subset of TA505. This group has been active since 2014 and is typically financially motivated. They were recently noted as the group responsible for the MOVEit Transfer Zero-Day exploit in May 2023. They're considered one of the largest, if not the largest, phishing and mouse spam distributors worldwide, estimated to have compromised over 3,000 US-based and over 8,000 global organizations. They have played many roles in the cybercrime community, including acting as a ransomware service operator and an affiliate with other ransomware. As a service operator, they have acted as an initial broker and as a customer of other initial access brokers, known as IABs, selling access to compromised corporate accounts. In other instances, they have also acted as a large botnet operation operator, specializing in financial fraud and phishing attacks. They are a significant player in the global cybercrime industry and have been a driver of many trends in the cybercriminal underworld.

TA505 targets education, finance, healthcare, hospitality, retail, and many other industries worldwide. They're known for their long-term cyberattack lifecycle: stealthily hiding in a target's network and conducting reconnaissance operations for months at a time, all while avoiding detection as they identify high-value targets in infrastructure. They typically use ransomware, known as lucky, as their primary cyberattack tool but have also been known to use other types of malware, including the CLOP ransomware, FlawedAmmyy, P2P RAT, Cobalt Strike, Mimikatz, and banking trojans like Dridex.

The next type of threat group is cybercriminal gangs that engage in illegal activities purely for financial gain. Many of their motivations include stealing sensitive data, conducting ransomware attacks, credit card fraud, credit card theft, identity theft, and then selling that information on the dark web. Cybercriminals often operate in organized groups or as individual actors. Some of these groups can be state-sponsored as well. They leverage a wide range of techniques, including phishing, malware, distribution, social engineering, and exploiting vulnerabilities in software and systems. They often target individuals, businesses, financial institutions, and other organizations to extract monetary value from their criminal activities.

Some well-known cybercriminal gangs include Dark Side, REvil, and Maze. These groups are known for carrying out high-profile ransomware attacks. The ramifications of cyber activity can include many things. Financial losses can occur. They primarily operate with the intention of financial gain. Cybercriminals' activities, such as ransomware attacks, data breaches, and identity theft, often result in significant financial losses for organizations across the globe.

Many businesses face costs associated with the response, including data recovery, legal fees, regulatory fines, and reputational damage, not to mention opportunity cost loss because of the actions these criminal gains commit. On an individual basis, individuals can suffer financial hardship related to funds being stolen from their accounts, fraudulent transactions committed against those accounts, and even identity theft.

Other activities include the targeting of sensitive personal information and financial information. These breaches can expose an individual's identity, fraud, and unauthorized access to their accounts. Loss of privacy against individuals erodes trust in online services and can have long-term effects on the individual and the organization to whom these criminal acts are committed.

In 2022, we witnessed a staggering number of victims hit by a breach: 422 million. This is up from the 294 million in 2021, and the cost of the breach continued to rise. Criminal gangs can also have the intent to disrupt infrastructure by creating distributed denial-of-service attacks, otherwise known as DDoS attacks. They can disrupt the functioning of critical infrastructure systems and often target government networks, financial institutions, healthcare facilities, utility companies, and many other companies and industries that operate in the critical infrastructure sectors. This can disrupt and cause several outages to services, cause financial instability, and even threaten public safety.

Criminal gangs often damage the reputation of the organizations that they commit these crimes against. This reputation damage can also include harm to the brand image. Data breaches and security incidents erode customer trust and confidence, leading to a loss of business, decreased shareholder value, and challenges in rebuilding a company's brand image. Criminal gangs often target intellectual property and steal this intellectual property, trade secrets, or proprietary information for their own gain. This can result in the loss of competitive advantage and hurt the market position of the company being targeted, not to mention the significant financial loss for the companies whose intellectual property has been stolen. It can also be sold to competitors or others to use in counterfeiting operations, which further damages the company's reputation and the financial losses that might occur.

There is a cost to dealing with these incidents. Criminal activities and the increasing prevalence of these activities have necessitated significant investments in cybersecurity measures and resources. Companies need to allocate substantial budgets and resources for security measures, employee training, incident response, and monitoring to try and thwart these attacks. These costs can be especially taxing on a smaller or medium-sized enterprise and lead to additional economic burdens, such as lost opportunity costs.

The impact of criminal gangs has far-reaching and broader economic impacts across both the United States and the globe. Cybercrime costs the global economy billions and billions of dollars annually. This decreases productivity, increases insurance premiums for all corporations, can even cause job losses, and obviously can cause consumer confidence erosion. Cybercriminals who often target social systems such as health care, education, and emergency services cause disruptions to these services and can directly impact public safety, patient care, or even educational opportunities. They often spread misinformation and use social engineering tactics to contribute to social unrest, political unrest, as well as erosion in institutions that are targeted by them.

Another threat group is known as hacktivists and ideologists. They combine hacking skills with political or ideological motives. They aim to promote social and political causes, raise awareness for a cause, or simply seek justice by targeting individuals, organizations, or governments that they perceive as oppressive, corrupt, or unethical. Hacktivists often employ various techniques that include web defense, defacement, website defacement and distributed denial-of-service, otherwise known as DDoS attacks, leaking sensitive data and causing disruption to information. Hacktivists often leverage social media platforms to disseminate messages and coordinate their hacking activities.

Some of the widely known hacktivist groups include Anonymous, the Lizard Squad, and the Lowes Act, which gained prominence for their hacktivist activities targeting entities such as the government, corporations, and many others. The term hacktivist is a combination of the words hacker and activist. Their motivation and approach to cyber activity typically employ techniques and digital activism to promote social or political causes.

Some of the key characteristics related to hacktivism are that they're driven again by their ideological and political motives, and they seek to advance those causes. Their activity is typically rooted in a desire for change rather than financial gain, and tactics typically include the defacement of websites, data leaks, or unauthorized access to systems. They target entities they perceive as acting against their causes or engaging in unethical behavior.

Hactivists typically disseminate messages and statements to raise public awareness about their causes. Those messages are usually disseminated on social media, websites, or other digital channels. Most often, hacktivist actions fall into a gray area in terms of legality and ethics. Some may argue that activists are engaging in civil disobedience or digital activism, and their actions can certainly be considered illegal under existing laws, especially if they involve unauthorized access and breaches. Hacktivist action can have significant consequences, sometimes positive and often negative.

While they have occasionally succeeded in drawing attention to important issues and sparking public debate, their activities often cause service disruption, include data breaches, and cause collateral damage to innocent victims. These ethical dilemmas of hacktivists are often a topic of debate and can vary depending on the perspective of the individual.

Insider threats are another type of group. This refers to individuals within an organization who misuse their authorized access to compromise the security of the organization. Typically, their motivations include financial gain, revenge, or some ideological belief. Insiders often have intimate knowledge of an organization's processes and certain vulnerabilities, which make them particularly nefarious. This threat group may exploit their privilege to steal sensitive data, sabotage systems, or leak highly sensitive data. Insider threats can be very challenging to detect due to the legitimate access they've been given and the trust principles we've given them, and then they abuse this trust.

Famous insider actors have included Edward Snowden, who leaked classified NSA documents, and Chelsea Manning, who also leaked classified information. Cybersecurity measures are frequently focused on threats from outside an organization rather than threats posed by trusted individuals inside the organization.

Insiders are a source of many losses to critical infrastructure industries. Well-publicized insiders have caused irreparable harm to national security interests. In fact, insider threats affect over 34% of businesses globally every year, and insider attack incidents have increased by 47% over the last few years, from 2021 to 2023.

Terrorist groups are also another type of threat actor. They're a significant concern for today's interconnected world, and these organizations recognize the potential of cyberspace as a platform for communication, recruitment, fundraising, and propaganda-type activities.

Terrorist groups engage in cyberattacks with the aim of causing disruption, spreading fear, or gaining strategic advantage. These attacks often target critical infrastructure, financial systems, or government networks. Consequences include service disruption, compromised data, financial loss, and even threats to public safety. These organizations often promote tactics on social media, including encrypted messages and applications, to exploit and disseminate them for extremist ideologies.

Terrorist groups recruit vulnerable individuals and also facilitate the planning of attacks. They often utilize cyberspace to spread their propaganda, recruit others, and generate support for their causes. They manipulate information, create fake news, and exploit social media to amplify their message. The dissemination of extremist content poses a threat to public safety, social cohesion, and the credibility of our online platforms. Not to mention, many of these channels provide terrorist groups with opportunities for fundraising, money laundering, and other illicit financial transactions. They often exploit cryptocurrencies, online payment systems, and crowdfunding sources to raise funds to support their terrorist activities.

Another group known as thrill seekers and trolls, sometimes called script kitties, refers to an individual who engages in cyber behavior to provoke an emotional response, incite a conflict, or spread misinformation. Many cyber trolls may not have the technical expertise of a hacker, such as a nation-state or even a cybercriminal, but their actions can contribute to cybersecurity challenges. They often spread false information or rumors through social media, other forms, and even in the comment section of many websites. This leads to inaccurate information, misleading guidance, or deceptive claims. Misinformation can cause confusion for individuals, making them vulnerable to additional cyber threats.

Trolls or thrill seekers often employ social engineering techniques to manipulate an individual into revealing sensitive information or performing actions that may compromise their security. This is often done by raising the sense of urgency, causing an emotional trigger, and deceiving people into disclosing this information. They often engage in online harassment and threaten to reveal private information about an individual.

Another type of threat group that is not often thought about is competitors. Competitors of a company can use various types of methods to achieve their motives. Competitors often target a company's intellectual property through cyberattacks to gain access to specific trade secrets, proprietary technology, or other research that may have high value. They often hack into networks, breach databases, and conduct inside attacks to steal this information. Sometimes, competitors may also commit sabotage against an organization. They can also perform distributed denial-of-service attacks that overload a website.

A competitor may often engage in cyber espionage to gather intelligence on their rivals, which can also involve targeted phishing campaigns, social engineering, advanced persistent threats, or APTs to infiltrate the target's networks. Many times, it's not just for financial gain; it might be brand or reputational damage. They often target the supply chains of their rivals to compromise the integrity of the product or service that their competitor may be utilizing. Often, competitors may use a fake website or phishing site to lure in their competitors. This aims to redirect legitimate customers to unauthorized accounts and gain sensitive information. It's important to note that attributing cyberattacks to specific corporate competitors can be challenging due to the nature of the cyber investigation and the potential for false-flag operations.

Many of these threat actors we've discussed can use various means across organizations. The threat actor can commit many types of attacks, ranging from supply chain attacks to ransomware attacks. Let's discuss a few of the techniques that these threat actors can use (we will discuss some of these more in Chapter 4).

1. The first and most commonly known cyberattack refers to *Malware*, which is malicious software, including viruses, worms, spyware, ransomware, and adware. Some malware disguises itself as legitimate software, known as a trojan virus. Ransomware is a

type of malware that blocks access to key infrastructure components. Spyware often steals confidential information without knowledge. Adware displays advertising content, such as banners, on a user's screen, which may cause a disruption. Malware breaches a network through a vulnerability, typically. When a user clicks on a dangerous link, it downloads an email attachment or when an infected pen drive is used.

2. *Phishing* attacks are another type of attack, one of the most prominent and widespread types of cyberattacks. Typically, it involves social engineering attacks, where the attacker impersonates a trusted contact and sends the victim malicious emails. The victim opens the emails and clicks on the malicious link or opens the malicious attachment. This allows the attacker to gain access to confidential information and account credentials. Then, they can install malware through this attack.

3. *Password* attack. This attack is where a hacker cracks your password with various programs and password-cracking tools such as John the Ripper, Ash Cat, Kane, Abel, or Aircrack. There are many different types of password attacks, such as brute force attacks, dictionary attacks, and keylogger attacks.

4. A *Man-In-The-Middle* attack, known as an eavesdropping attack, is where an attacker comes in between a two-party communication application, i.e., between the session between a client and a host. This allows the hacker to steal and manipulate data. Another type of attack is a SQL injection attack. This attack occurs on a database-driven website where the hacker manipulates a standard SQL query. It is carried out by injecting malicious code into a vulnerable website search box, thereby

making the server reveal crucial information. This results in the hacker being able to view, edit, or delete tables in the database. Hackers can also gain administrator rights through this type of attack.

5. *Denial-Of-Service (DoS)* attacks cause significant threats to companies. These types of attacks target system servers and networks and flood them with traffic to exhaust their resources and the bandwidth they use. When this happens, incoming requests become overwhelming for the servers, which results in the website or host being unable to process additional requests, leaving legitimate service requests unattended. This is also known as a DDoS attack, or distributed denial-of-service attack, when the attacker uses multiple compromised systems to launch this type of attack.

6. Another type of attack that is widely used is a *Business Email Compromise* (BEC). This targets businesses and organizations by using email. The attackers impersonate a trusted source to trick the victim into transferring funds or sensitive information to the attacker. This is usually done by raising the sense of urgency or using a position of power.

7. *Zero-Day Exploits* are attacks that happen after the announcement of a network vulnerability, and there is no solution for this vulnerability in most of these cases. Once the vulnerability occurs, the vendor may notify the public to be aware of this vulnerability. This also informs attackers. Mitigating this type of vulnerability may take vendors multiple weeks or months to remediate. This gives attackers time to target the disclosed

vulnerability. Attackers then exploit this vulnerability before a patch or solution is implemented.

8. *Supply Chain* attacks have made the news recently due to a number of high-profile attacks. These attacks often target software or hardware supply chain vulnerabilities, much like we saw in the SolarWinds attack in 2020 to collect sensitive information. They may use other methods in their attack, but it is meant to affect a supply chain that can have an effect on many companies around the globe.

There are many other types of attacks threat actors use, such as Spoofing, Code Injection, various types of Phishing, Social Engineering, and Advanced Persistent Threat attacks (APT). The list is very lengthy.

The cybersecurity landscape is fraught with diverse types of cyber threat actors, each with unique motivations, capabilities, and tactics. State-sponsored hackers, cybercriminals, hacktivists, and insider threats pose significant risks to individuals, organizations, and governments. Understanding these actors is crucial for developing strong cybersecurity strategies, implementing effective countermeasures, and promoting a resilient digital environment.

In 2022, external notifications were more prevalent as a notification source, regardless of the investigation type. Intrusions related to ransomware were notified by an external entity in 70% of investigations. Continuous, vigilant international cooperation and advancements in cybersecurity technology are vital for mitigating the threat posed by these threat actors and safeguarding digital infrastructure around the globe.

Many of the techniques these threat actors use and various types of attacks are more sophisticated. Individuals must rely on many of their own defenses, including locking their credit files. Organizations have to use a variety of security tools to thwart these attacks. These attacks seem to continue to grow in scope and sophistication. It requires all of us to be intelligent to defend against these attacks.

CHAPTER 4

Threat Actor Tactics, Techniques, and Procedures

Cyber warfare techniques have been going on for decades. They were just not publicized like today. One of the first known events related to technology, as discussed previously, was the Soviet Union Siberia pipeline explosion. U.S. President Ronald Reagan approved that particular CIA plan in 1982 when the Soviet Union asked the U.S. to assist with software related to a Siberian natural gas pipeline. The incident is described in detail in the book *At the Abyss, an Insider's History of the Cold War*, written by Thomas C. Reed, a former U.S. Air Force Secretary serving in the National Security Council at the time of the incident.

Cyberattacks have certainly escalated since 1982. Today, attacks happen over 4,000 times daily, quite different from 40 years ago. Statistics from the Department of Homeland Security, Cybersecurity and Infrastructure Security Agency, known as CISA, state that 600,000 Facebook accounts are hacked every day. Cyber threat actors are employing various techniques to target global critical infrastructure, aiming to disrupt essential services, compromise sensitive systems, or gather intelligence. Here are some of the common techniques observed in attacks on critical infrastructure:

Attackers often use phishing emails, deceptive websites, or social engineering tactics to trick employees of critical infrastructure organizations into revealing login credentials or downloading malicious attachments. These techniques allow attackers to gain unauthorized access to systems or deliver malware to compromise infrastructure networks.

Ransomware attacks have increasingly targeted critical infrastructure. Attackers exploit vulnerabilities, gain initial access through phishing or compromise credentials, then deploy ransomware to encrypt systems and demand a ransom for their release. These attacks disrupt operations and lead to significant financial losses.

Threat actors may also compromise the supply chain of critical infrastructure systems by injecting malicious code, hardware backdoors, or software or hardware updates. This enables them to gain unauthorized access to or control the critical systems.

Cyber threat actors sometimes discover and exploit previously known vulnerabilities in software. These are known as zero-day exploits. When the cyber threat actor leverages these vulnerabilities before they are patched, they can infiltrate critical infrastructure systems and gain control or extract sensitive information.

Distributed denial-of-service (DDoS) attacks are commonly used to overwhelm real, critical infrastructure networks, rendering the services unavailable. Attackers employ botnets or other means to flood the networks with a high volume of traffic, which exhausts the resources of infrastructure networks and causes service disruption.

Data and ICS, otherwise known as supervisory control, data acquisition, and industrial control systems used in critical infrastructure, are often targeted. Attackers exploit vulnerabilities in these systems to gain unauthorized access, manipulate processes, disrupt operations, or cause physical damage. This can even increase the risk to the physical safety of those who operate this equipment. Insiders with privileged access to critical infrastructure systems also pose a significant risk. They may willingly or unknowingly facilitate cyberattacks by leaking sensitive information, misusing credentials, or intentionally disrupting operations.

Another attack often observed on critical infrastructure is an APT attack, otherwise known as advanced persistent threats. These attacks are often associated with state-sponsored actors. These attacks involve a prolonged presence within a network, stealthy movement, and continuous data exfiltration. APTs target critical infrastructure to gain access to sensitive information, disrupt services, or conduct surveillance. Some attacks on critical infrastructure involve physical elements combined with cyber components.

Attackers may use cyber techniques to gain unauthorized access to physical control systems, enabling them to manipulate or sabotage critical infrastructure, which can also cause harm to humans if there is catastrophic damage done. Entire books have been written on how threat actors are exploiting our critical infrastructure in a variety of ways. It is only my intention to scratch the surface of these attack methods so that it may pique one's curiosity to learn more or discuss defenses within their organizations and with their security team.

Exploits are done in many different ways. Tactics used by cyber threat actors to exploit human behavior, including phishing, spear phishing, or other methods, are known as social engineering. This is a technique used by cyber threat actors to manipulate and deceive individuals into divulging sensitive information, performing certain actions, or granting unauthorized access to systems or resources. Social engineering attacks exploit the inherent vulnerabilities of individuals rather than the targeted technical vulnerabilities.

Social engineering is exploited in cyberattacks by phishing, pretexting, baiting, tailgating or impersonation, spear phishing, watering hole attacks, quid pro quo, or social media exploitation.

Let's discuss each:

1. *Phishing* attacks involve sending deceptive emails, messages, or phone calls that appear to be from a legitimate source, such as a trusted organization or college. Attackers often create a sense of urgency or fear, enticing recipients to click on malicious links, download malware or infected attachments, or provide login credentials and personal information on fake websites.

2. *Pretexting* involves creating a fictional scenario as a pretext to manipulate individuals into revealing information or performing actions they normally would. Attackers may use a person in an authoritative position, such as a company executive IT support personnel or a trusted vendor, to gain the target's trust and convince them to disclose sensitive data or grant unauthorized access.

3. *Baiting* attacks lure victims with the promise of something desirable, such as a free download, a discount, or a prize. Attackers may distribute infected emails or create enticing download links where individuals will take the bait. Once the victim interacts with the bait, malware is deployed, compromising the system.

4. *Social Engineering* attack: physical tailgating or impersonation. Individuals may impersonate an employee, a contractor, or a service provider to gain unauthorized access to restricted areas. This often involves following a legitimate person closely or using fake identification to bypass security measures and gain entry to a secure facility or network.

5. *Spear Phishing* is a targeted form of phishing where attackers gather information about their victims, such as their names, job roles, or affiliations, to personalize the attack. A well-crafted message or call that appears to be from a trusted source or known colleague can increase the chances of success for the threat actor by convincing the victim to disclose sensitive information or perform specific actions.

6. *Watering Hole* attacks target websites or online platforms that are frequently visited by targeted individuals or organizations. Attackers compromise these trusted websites by injecting malicious code or exploiting vulnerabilities. When an individual visits the compromised site, their systems can be infected with malware, allowing the attacker to gain access to the system and/or their credentials. For example, an industry group might be targeted. By injecting malicious code or exploiting vulnerabilities in these websites, the attackers can deliver malware and ransomware to the individual's device. This attack targets a specific group or industry, increasing the likelihood of a successful attack.

7. *Quid Pro Quo* attacks often promise a benefit or service in exchange for the victim's sensitive information. An attacker may pose as a technical support representative and offer assistance

with a reported computer issue. In return, the victim is persuaded to provide login credentials or other confidential information. In a social media exploit attack, the attacker leverages information available on social media platforms to gain personal details about their targets. This information can be used to personalize social engineering attacks, such as creating a fake relationship, impersonating a friend or colleague, or gathering other details to answer security questions. To defend against social engineering attacks, individuals and organizations should implement security awareness training programs, verify requests for information or actions, and maintain strong cybersecurity hygiene.

8. *Malware and Ransomware* attacks and the impact on businesses and organizations they cause have increased dramatically over the last few years. Malware and ransomware attacks are commonly exploited in cyberattacks to gain unauthorized access, compromise systems, steal data, and extort a victim. These malicious programs are often designed to infiltrate and disrupt computer systems or networks. The following methods are techniques by which malware and ransomware are exploited in a cyberattack.

Attackers often distribute malware or ransomware through email attachments or embedded links. Victims may unknowingly download and type in files or click on malicious links, which then execute the malware on their system. These emails often employ social engineering techniques to convince recipients to open the attachment or click on the links. Malicious code can also be injected into a legitimate website or compromised web pages. When users visit these websites, the code automatically downloads and executes the malware on their devices without their knowledge or consent. Exploit kits are often used to target vulnerabilities in web browsers, plugins, or operating systems.

Attackers may also place malicious advertisements, known as "malvertisements," on legitimate websites or ad networks. When users click on these ads or even visit websites hosting the malvertisements, they are redirected to websites hosting malware. They may not even know that they have been directed to an illegitimate or malicious website. The malware is then downloaded onto the user's device, compromising its security.

Cybercriminals often distribute malware through file-sharing networks or websites that offer pirated software, games, or media. Unsuspecting users who download and install these files may inadvertently infect their systems with malware or ransomware. They may then also pass this on to their friends or colleagues, further infecting other systems. Malware and ransomware can exploit vulnerabilities in software operating systems or network devices.

Attackers continuously search for and exploit security fraud to gain unauthorized access. Once a vulnerability is identified, they use it to deploy the malware or ransomware onto the targeted system. Attackers may also compromise the supply chain, be it software or hardware, by injecting malware or ransomware into legitimate software updates or installation packages that may be used. When users download and install these compromised updates, their systems become infected with malicious code. Once malware or ransomware is successfully executed, it can carry out various malicious activities such as data exfiltration, system takeover, encryption of files, or surveillance. Ransomware specifically encrypts files on the victim's system and demands a ransom payment in exchange for the decryption key.

To defend against malware and ransomware attacks, individuals and organizations should employ security best practices, including using reputable antivirus or anti-malware software, regularly updating software and operating systems, and exercising caution when opening email attachments or clicking on links. Avoid downloading files from untrusted sources, and implement very strong backup and recovery processes. Additionally, user security awareness training, network segmentation, and proactive vulnerability management can aid in mitigating the risk of malware and ransomware attacks.

Now, I'd like to analyze the vulnerabilities of supply chains and the potential for attackers to compromise systems through third-party vendors and suppliers and explore strategies for securing the supply chain.

A supply chain cyberattack, also known as a software supply chain attack or value chain attack, targets the software or hardware supply chain to compromise the security of a target organization or customers. In this type of attack, rather than directly targeting the organization's own systems, the attacker infiltrates and exploits the trusted relationships and dependencies within the supply chain to gain unauthorized access, introduce malicious code, or compromise the integrity of the products or services being delivered.

The following are some key characteristics and examples of supply chain cyberattacks:

- Attackers will often infiltrate the software update process of a trusted vendor or supplier and introduce malicious code into a legitimate software update. When users or organizations download and install these compromised updates, they unknowingly introduce malware backdoors into their systems.

- Another tactic attackers use in a supply chain attack is targeting the hardware supply chain by tampering with the manufacturing, assembly, or distribution process. This may involve inserting malicious components or modifying legitimate hardware to create vulnerabilities or enable unauthorized access to the target systems.

- Attackers will also target the dependencies or the libraries used in software development. The attacker can impact multiple software applications or systems that rely on components by compromising or manipulating a widely used library or component. This technique allows for the widespread compromise or exploitation of the supply chain.

- Attackers often insert themselves in the software development process of a trusted vendor or supplier, gaining access to source code repositories or build systems. By tampering with the development environment, the attacker can inject malicious code into the software or introduce vulnerabilities that can be exploited at a later stage.

- Attackers often introduce counterfeit or compromised hardware components into the supply chain. These components may have hidden vulnerabilities or backdoors that can then be exploited to compromise the security or integrity of the systems that use them.

The impact of a supply chain cyberattack can be severe, with widespread consequences for the target organization and its customers, such as those we've seen in the SolarWinds attack. Attacks like this can lead to unauthorized access to sensitive information, compromise the integrity of systems or software, enable further attacks, or facilitate espionage or sabotage.

Defending against supply chain cyberattacks requires a comprehensive and multifaceted approach. Organizations should conduct a thorough risk assessment of their supply chain. Additional measures can include scrutinizing and monitoring vendors and suppliers for security practices, implementing secure development and verification processes, establishing strong vendor management and oversight protocols, and continuously monitoring infrastructure for signs of compromise or tampering.

Supply chain resilience and contingency plans should also be developed to minimize the impact of potential attacks and ensure business continuity. Collaboration and information sharing within industry sectors can also help identify and address potential threats in the supply chain.

Now, let's briefly examine the weaknesses in cloud security and the challenges of security data and systems in the cloud. Additionally, we will discuss weaknesses in IoT, the Internet of Things, and the associated security of IoT devices. Vulnerabilities in each of these can have devastating effects on an organization.

Cloud security weaknesses include a weak authentication mechanism, improper access controls, or misconfigured permissions that can allow unauthorized individuals to gain access to sensitive data or cloud resources. Cybercriminals can target cloud providers, leading to data breaches and unauthorized access to customer data. Weak encryption or improper data handling practices can exacerbate these risks. Malicious insiders with privileged access to cloud environments can abuse their privileges to steal or manipulate data, disrupt services, or introduce malware. Another vector for attacking cloud systems is application programming interfaces that connect cloud services and cloud client applications that may have vulnerabilities that can be exploited by an attacker.

Weak authentication, improper input validation, and insufficient encryption can be exploited to gain unauthorized access or manipulate data. Additionally, there is a risk of cross-tenant data leakage or attacks in multi-tenant cloud environments where multiple users share the same infrastructure. Insufficient isolation between tenants can lead to unauthorized access or information disclosure.

IoT (Internet of Things) security weaknesses include insecure communications, a lack of device management and updates, default or weak credentials, a lack of physical security, inadequate data protection, and a lack of standardization. First, we'll discuss insecure communication. IoT devices, or Internet of Things devices, often rely on insecure communication protocols such as Wi-Fi or Bluetooth, which can be intercepted or tampered with with inadequate encryption or weak authentication mechanisms that can enable unauthorized access or data interception. Many IoT devices lack proper mechanisms for remote management and updates. This leaves them vulnerable to known vulnerabilities and exploits, as patching or firmware updates may not be feasible or straightforward. IoT devices also often come with default credentials that are rarely changed by users. Attackers can easily guess or exploit these default or weak credentials to gain unauthorized access to the device or the network.

Additionally, IoT devices deployed in physical environments may lack adequate physical security measures. This makes them susceptible to physical tampering, unauthorized access, or theft, which can compromise their integrity or allow for the extraction of sensitive data. IoT devices often collect and process sensitive data. However, data encryption, proper storage, and secure data transmission are not always implemented, leaving the data vulnerable to interception or unauthorized access. The lack of standardized security measures across different IoT devices and manufacturers can also lead to inconsistent security practices and make it challenging to ensure a consistent security posture across the IoT ecosystem. Another weakness in IoT security is that many devices do not allow an agent to be installed on them that could help protect them.

Addressing these weaknesses requires a combination of technical measures, industry standards, and user education. Cloud providers and IoT device manufacturers should implement robust authentication mechanisms, encryption protocols, and secure coding practices.

In the U.S., we are starting to see initiatives around accountability related to manufacturers. Regular security updates and patch management should be prioritized. Users and organizations should also take the initiative to secure their cloud environment and IoT devices by using strong, unique passwords, disabling unnecessary features, regularly updating their hardware, and monitoring for suspicious activity.

Additionally, industry collaboration, regulatory frameworks, and standards development are essential to driving consistent security practices and ensuring the resilience of the cloud and IoT ecosystems.

Additional cyberattack vectors include insider threats, as discussed in the previous chapter. This refers to the risk of an employee or unauthorized individual misusing their access to company data or systems. This can include intentional data theft or sabotage, accidental data leaks, or negligent behavior that puts the company's security at risk.

Distributed denial-of-service attacks are another type of cyberattack that involves overwhelming a system or network with traffic, making it unavailable to users. These attacks can cause significant disruption to a company's operating operations and be difficult to mitigate.

I suppose since you're reading a book on cyber warfare, you are interested in knowing how to defend against these attacks. Step one in defense is to know how they work. Once one better understands how they work, they are more prepared to defend against them.

Attacks carried out by a skilled adversary may consist of multiple stages that are repeated. They are persistent in their methods and have plenty of time to carry out their attack. Not only that, they also only have to be right once, whereas defenders, including the technology deployed, have to be right 100% of the time. That is why detection, response, and recovery are so important. It will happen to you or your organization if one stays in the cyber game for a long time. Targeted attacks and untargeted attacks work differently. In an untargeted attack, sometimes known as spray and pray, attackers discriminately attack as many devices, resources, services, or users as possible. The victim is not important; only the motive, the success rate, and the number of systems or resources with vulnerabilities are.

Techniques threat actors may deploy include phishing, watering, hauling, ransomware, scanning, and other methods. A targeted cyberattack is just that: a victim, whether an organization, individual, or group of entities, has been singled out due to an interest, or they have been paid to target these entities. Threat actors will take months to perform reconnaissance, quietly prodding their intended victim in an attempt to find the best method to exploit an organization and/or user system.

Techniques threat actors may deploy include spear phishing, deploying botnets, supply chain attacks, and several others. The idea is to target your organization's systems, services, or personnel in the office and sometimes at home. They want to cause the most damage possible. How do they do this? There is a widely known and accepted framework by cyber practitioners that was developed by Lockheed Martin called the cyber kill chain framework. It consists of seven steps: reconnaissance, weaponization, delivery, exploitation, installation, command and control, and actions on objectives. Let us simplify this a bit and discuss four steps. Investigation. Attackers will use any means available to find any vulnerability to take advantage of an attempt to exploit.

Attackers will use social media and publicly available information such as domain names, websites, and the Dark Web to gather intel on a target. Attackers will utilize readily available information, toolkits and techniques, and commercialized network scanning tools to collect and assess any information about your organization's systems and personnel. Attackers will even use other threat actor services to perform this stage. The next stage is the delivery stage. Throughout the delivery stage, the attacker will look to pivot where they can exploit a vulnerability they have discovered or one that may exist. They may access an organization's online services, send an email with malicious links or web addresses, or create fictitious websites to draw employees or individuals in. Attackers are looking for the easiest path into critical infrastructure that will allow them to cause the most damage. The breach stage's success may depend on the threat actor's motives. They may disrupt an organization's systems by making changes affecting operations or gaining access to intellectual property or privacy data, such as personally identifiable information. They may cause physical damage to systems that disrupt operations, such as what we saw in the Stuxnet attack many years ago, or they may simply deface a public website as part of their activity.

Often, these attacks are for monetary gain or shortcuts for those who need to innovate on their own. The next stage is the impact. An attacker may seek to exploit. An attacker may seek to explore your systems, expand their access, and establish a persistent presence, a process sometimes called consolidation. With administration access to just one system, they can try to install automated scanning tools to discover more about your network and take control of more systems. Determined and undetected attackers continue until they have achieved their end goal, such as retrieving sensitive information or creating payments into a bank account. They have access to disrupting operations or politically motivated actions. These attackers will try to leave undetected, create additional holes they can exploit at a later date, and/or destroy systems for publicity or to raise their image among other hackers.

There are several emerging technologies that are having a significant impact on cybersecurity and cyber warfare. These technologies can be exploited by threat actors or used to defend against attacks, both presenting new challenges and offering potential solutions. Here are some of the key emerging technologies shaping the cybersecurity landscape now and into the future:

Artificial intelligence and machine learning technologies are being utilized in cybersecurity to enhance threat detection, automate security operations, and analyze vast amounts of data for anomaly detection. These technologies can improve the efficiency and accuracy of security systems, identify previously unknown threats, and enable real-time responses to cyber actors. Emerging technologies can also be used by threat actors for their own gain.

Quantum computing has the potential to revolutionize both offensive and defensive cyber capabilities. While quantum computing may threaten existing encryption algorithms, quantum-resistant cryptography is being developed to safeguard sensitive data in the post-quantum computing era.

The rapid proliferation of the Internet of Things introduces new vulnerabilities and expands attacks. Security concerns arise due to the insufficient security measures implemented on IoT devices. Securing IoT infrastructure requires robust authentication, encryption, and access controls to prevent unauthorized access and protect against IoT-based bots.

Blockchain technology, known for its application in cryptocurrencies, offers potential benefits for cybersecurity. Its decentralized and immutable nature can enhance data integrity, facilitate secure transactions, and enable secure identity management. Blockchain-based solutions are being explored for secure data sharing, supply chain integrity, and secure decentralized systems.

Cloud computing has transformed the IT landscape, but it also introduces new security challenges. Secure cloud infrastructure, encryption access controls, and data protection measures are essential to mitigate the risk associated with shared resources, data breaches, and unauthorized access. Biometrics and behavior analytics are also modern technologies that are enhancing cybersecurity. Biometric authentication methods such as fingerprints, facial recognition, or iris scans provide enhanced security for user authentication.

Behavioral analytics analyzes users' behavior patterns to detect anomalies and potential insider threats, enhancing user authentication and access control mechanisms. The rise of autonomous systems and robotics presents new security challenges. Securing these systems against hacking, tampering, or malicious control is critical in preventing physical harm or disruption caused by compromised autonomous technologies.

Another emerging technology is big data analytics. Analyzing and processing large volumes of data enables better threat detection and predictive analytics. Big data analytics can identify patterns, detect anomalies, and provide actionable insight to strengthen cyber defenses. A threat actor can also use this data to attack someone who is particularly susceptible to specific attacks.

Other emerging technologies include software-defined networks, which enable more centralized management and control. Edge computing, which brings data storage closer to the source, reducing latency and bandwidth requirements, and secure access service edge, known as SASE, is a cloud architecture model that combines network and SECaaS (security as a service) functions together and delivers them as a single cloud service.

While these emerging technologies bring advancements, they also introduce novel cybersecurity risks. These technologies, if used in harmful ways, have the means to escalate, increasing the likelihood of a cyber war. It is essential for organizations and decision-makers to stay informed, adopt security practices, develop regulatory standards, and address the evolving cybersecurity landscape.

Collaborative efforts among industry, academia, and government are crucial to harnessing the potential of these technologies while ensuring a secure and resilient cyber environment. To protect global infrastructure, organizations and governments should implement robust cybersecurity measures such as network segmentation, access controls, intrusion detection systems, and regular patching.

Continuous monitoring, threat intelligence sharing, and incident response plans are also crucial to detecting and responding to attacks effectively. Additionally, collaboration between the public and private sectors is essential to strengthening the overall security posture and resilience of critical infrastructure.

We have discussed many threat actors and the types of attacks that they commit. It is crucial that we all learn about our adversaries and learn how to protect ourselves against these attacks to be better prepared now and in the future.

From the Desk of a CISO: A CISO's Perspective

Critical Infrastructure Risks and Defenses

featuring Clif Triplett

Clif Triplett Bio: Clif Triplett is the Executive Director for Cybersecurity and Risk Management at Kearney. He has in-depth knowledge of information technology and global experience spanning the government, oilfield services, tractor, automotive, aerospace, defense, and telecommunications sectors. In 2016, Clif worked with the Obama administration to address the cybersecurity challenges brought to light by the US Office of Personnel Management (OPM) data breach, in which 21.5 million individuals' background investigation records were compromised. He is currently working with a Middle Eastern government to revise and standardize national practices and laws related to data classification and the protection requirements for classified information.

Clif's notable engagements include:

- Establishing a process for evaluating new cyber technologies in the U.S. federal government

- Establishing a cybersecurity compensation, recognition, and retention strategy for the U.S. federal government

- Consulting with the Italian government on developing an accelerated cybersecurity risk mitigation strategy

- Working with the National Security Agency, Defense Information Systems Agency, and the Department of Homeland Security (DHS) in the United States on cybersecurity architecture and priorities

- Working with the National Security Council to establish the basis for the U.S. cyber incident response process model

- Leading the first and only agency to fully implement the DHS cybersecurity initiatives within the Obama administration

- Leading a classified panel consisting of the U.S. intelligence agencies, Cyber Command, and DHS that represented the executive branch on the U.S. response to cyber threats

- Defining a defense-in-depth architecture and implementing a roadmap for cybersecurity that was implemented at OPM and influential in DHS programs

- Driving changes to the Federal Acquisition Regulation related to procuring IT and cyber technologies

- Co-chairing a council of the major healthcare insurance providers for the U.S. government that examined cyber breach reporting, risk mitigation, best practices, and Office of Inspector General assessment criteria and expectations

- Working with the Office of the President, Office of Management and Budget, Government Services Agency, and OPM on a strategy to deploy an alternative government identifier across the U.S. federal government, replacing the use of Social Security numbers in a secure and consistent manner.

Before joining Kearney, Clif served as an officer in the U.S. Army Signal Corps, working on top-secret and other advanced technologies. He has also served as a chief information officer and business unit leader at companies such as General Motors, Baker Hughes, and Motorola.

"The truth is that you always know the right thing to do. The tough part is doing it."
– Gen. Norman Schwarzkopf, United States Army General

"Start where you are. Distant fields always look greener, but opportunity lies right where you are. Take advantage of every opportunity of service."
– Robert Collier, American author

Securing the Critical Infrastructure Software Supply Chain

Today, software lies at the core of almost every critical infrastructure component that operates in our modern civilization. From power grids and transportation systems to healthcare and financial services, software-enabled technologies have become indispensable to the functioning of societies, especially in technologically advanced nations like the United States and many of our allies. Given software's ubiquitous role, ensuring its security, both from the perspective of the code that is developed and the supply chain ecosystem, is not merely a technical concern but a matter of national security, physical safety, and ensuring that our critical infrastructure maintains operability.

"A 'software bill of materials' (SBOM) has emerged as a key building block in software security and supply chain risk management," - CISA, https://www.cisa.gov/sbom

"If you take a look at secure coding, it can give you a false sense of security. Because what percentage of the operating system's operating code are you writing?" - Clif Triplett

"When you look at, 'Am I in a healthy place?' I have to look at, 'Do I really understand the software I am operating in the context of my enterprise?'" - Clif Triplett

Over the past few decades, we have seen that America's and other modern countries' critical infrastructure has become increasingly dependent on software to operate. Though this transition has enabled unparalleled efficiencies and capabilities, it has also introduced enormous numbers of vulnerabilities. Insecure coding can lead to system failures, unauthorized data access, and even remote control of vital systems by malicious threat actors. Software flaws in a power grid management system can be exploited to cause blackouts, impacting millions of citizens and the economy. The same impact can be said for the companies that operate critical infrastructure. We witnessed the Colonial Pipeline attack, which had a ripple effect across the United States, and President Joe Biden released an emergency declaration due to the attack.

"I think there has to be more energy in the integration of security coding, SBOM management, and trust scoring/asset management." - Clif Triplett

As technology advances, so does the threat landscape. This has been increasingly evident for the past few years as malware and hacking tools, including outsourced services, have become readily available on the dark web. The threat landscape has evolved dramatically. Nation-states, cybercriminals, state-sponsored hacking groups, and terrorist groups now employ highly sophisticated tools and techniques to cause great harm to the U.S. and global critical infrastructure. These adversaries recognize the value of targeting the software that powers critical infrastructures by exploiting coding vulnerabilities or compromising the software supply chain. Successful breaches can result in financial losses and threats to public safety, national security, and America's standing in the world.

"It really comes back to the NIST CSF and the identify phase. Have you really identified which code you need to pay attention to because it is really difficult to manage it all? Find out which components are critical to your infrastructure." - Clif Triplett

As mentioned previously, the software supply chain encompasses everything from the components that make up a software product to the processes used for its development, distribution, and maintenance. An attack on any point in this chain can compromise the end product. One example we discussed earlier was SolarWinds. The SolarWinds attack demonstrated how when attackers compromise a trusted software update mechanism to deploy malware to thousands of organizations, including governmental entities, the impact can be far-reaching and very devastating. Such vulnerabilities in the supply chain underscore the importance of a comprehensive approach to software security, going beyond just code review.

"When it comes to privileged access management, my objective is always zero privileged accounts in production without change control. If you had one show up and it was not authorized by change control, you could immediately shut it down." - Clif Triplett

Beyond just the direct security threats and impacts of those threats, vulnerabilities in software and its supply chain have significant economic and geopolitical ramifications. Significant breaches erode trust in American and global technology companies, diminishing their global market share. Another concern is that if foreign entities control or influence parts of the software supply chain, it raises red flags about espionage, data integrity, and the potential to sabotage the software supply chain.

"I think we will see more on the health side. Eventually, you will have to go through a lab to have your product certified by some semi-government agency, and that agency will also be heavily scrutinized. Like 800-53 level 2." - Clif Triplett

Waiting for a catastrophe to underscore the importance of secure software coding and supply chain security is not an option. Forward-thinking measures, such as adopting secure coding practices, continuous monitoring, third-party audits, and ensuring transparency in the software supply chain, are essential. National initiatives that set standards, provide guidance, and even incentivize secure practices can significantly improve the security posture of the United States and our allies, critical infrastructure to ensure each area of the supply chain is cyber-resilient.

"Are you really looking at your threat landscape? Are you thinking about how many people you give access to, how long they have access for, and have you done the appropriate department and network segmentation so you don't have lateral movement issues?" - Clif Triplett

There is no question that the United States and global critical infrastructure are becoming more intertwined with software. The imperatives of secure coding and supply chain security cannot be overstated. Protecting this infrastructure is analogous to safeguarding the nation's and globe's very heartbeat. Due to the ever-evolving threat landscape and the paramount importance of critical infrastructure, a focused, collaborative, and continuous effort is required from all stakeholders—government, private sector, academia, and citizens—to ensure that the digital foundation of the United States and our global partners remains secure and resilient.

CHAPTER 5

Mitigating Risks and Hacking Back

Critical infrastructure companies are essential to the functioning of our society. Companies in the 16 critical infrastructure sectors provide us with the power, water, transportation, and communication needed to live and work. Many of them are among the largest companies in the world. Companies from Fortune 500 to small local businesses can also be susceptible to cyberattacks. Each of these companies is increasingly vulnerable to cyberattacks. In recent years, several high-profile cyberattacks have occurred on critical infrastructure companies.

In 2017, the NotPetya cyberattack crippled the Ukrainian power grid. In 2018, the WannaCry ransomware attack infected hundreds of thousands of computers worldwide, including those of critical infrastructure companies in the United States and across the globe. In late 2020, the SolarWinds breach affected hundreds of global companies. In 2021, the Colonial Pipeline crippled U.S. gas supply along the East Coast and Gulf states of America. And it is not just the United States. In December 2022, a German multinational engineering and steel company, ThyssenKrupp, was hit by a major cyberattack. In June 2023, a pro-Russian hacktivist group attacked several European banking institutions, including the European Investment Bank, in retaliation for Europe's continued support of Ukraine.

These types of attacks are happening across the globe. No country, no government, and no company is immune to cyberattacks. These attacks have shown that critical infrastructure companies are not immune to cyberattacks and are increasingly becoming soft targets. And as the threat landscape continues to evolve, it is more important than ever for these companies to take steps to protect themselves. So, what should companies do? There are several strategies and defenses companies can undertake, from fundamental security, utilizing well-known cybersecurity frameworks, deploying advanced architectures such as zero trust architecture and incorporating robust network segmentation, implementing advanced technologies where possible, and planning for a post-quantum computing world, incorporating cyber threat intelligence and threat hunting, proper incident response planning, monitoring third and fourth party security posture, and even deploying attack simulations and robust user security awareness, along with collaborative efforts, are all good things to deploy. Let us drive home the point of fundamental security. There are many fundamental cybersecurity measures that critical infrastructure companies and other companies can take to mitigate the risk. Some of these measures include implementing strong cybersecurity policies and procedures. This includes things like having a strong password policy, requiring multi-factor authentication, and regularly updating software.

Educating employees about cybersecurity risks is another part of fundamental security. Employees are often the weakest link in the cybersecurity chain. By educating your employees about cybersecurity risks, regularly training them, and also utilizing phishing simulations or other simulations to test employees, companies can help prevent them from making mistakes that could lead to a cyberattack. Another fundamental exercise is to monitor for suspicious activity. Companies should use security tools to monitor their networks for suspicious activity or anomalous behavior. This can help detect and respond to cyberattacks early in the attack cycle. Having a plan for responding to cyberattacks is also particularly important. And it is fundamental that every company has a cyberattack plan or an incident response plan, regardless of their size.

Obviously, those plans will vary depending on the size and complexity of the company. However, every company should have one. In the event of a cyberattack, companies need to have a plan for how they will respond. This plan should include things like how they will contain the attack, how they will restore their systems, and how they will communicate with all stakeholders, internal and external. We will talk a little bit more about each of these in future discussions. Critical infrastructure companies can help mitigate the risk of cyberattacks by taking these fundamental cybersecurity measures.

In addition to the fundamental cybersecurity measures listed above, critical companies can take several other steps to improve their cybersecurity posture and mitigate cybersecurity risk. These steps include:

- *segmenting their network*, which involves dividing the network into different segments, each with its own security controls. This can help contain the damage in the event of a cyberattack by limiting the blast radius.

- They can also use *Intrusion Detection and Prevention Systems*, otherwise known as IDS and IPS systems. These systems can help detect and block malicious traffic. Backing up their data regularly, whether it's to an on-premises location or the cloud, will help minimize the impact of a cyberattack if data is compromised. Then, assess those backups regularly to ensure that they work as planned.

- Companies can also *Evaluate Their Cybersecurity Systems* regularly. This helps identify and fix any vulnerabilities.

By taking these additional steps, critical infrastructure companies can further reduce the risk of cyberattacks.

Cybersecurity is an ongoing process. As the threat landscape continues to evolve, critical infrastructure companies must constantly evaluate their cybersecurity posture and make changes as needed. By taking the necessary steps to protect themselves, critical infrastructure companies can help ensure they remain safe and secure. In addition to the technical measures listed above, critical infrastructure companies should also focus on building a culture of cybersecurity within an organization. This includes things like creating a sense of urgency around cybersecurity, rewarding employees for good security behavior, and holding employees accountable for security breaches.

By building a culture of cybersecurity, critical infrastructure companies can help to ensure that everyone in the organization is aware of the risk and is taking steps to protect the company from cyberattacks. Additional measures to mitigate risk include utilizing well-known cybersecurity frameworks. Cybersecurity frameworks are a set of best practices that organizations can follow to improve their cybersecurity posture. They provide a structured approach to identifying, assessing, and mitigating cybersecurity risks.

There are many benefits to incorporating cybersecurity frameworks into an information security program. Some of the key benefits include:

- *Increased Visibility and Understanding of Cybersecurity Risks.* Cybersecurity frameworks can help organizations identify and understand the risks they face. This information can be used to prioritize security investments and make better decisions about how to protect the organization.

- An additional benefit is an *Improved Cybersecurity Posture.* Cybersecurity frameworks can help organizations implement effective security controls. This can help reduce the likelihood and impact of cybersecurity attacks.

- *Increased Compliance With Regulations* is also a benefit. Frameworks can help organizations demonstrate compliance with relevant regulations. This can help protect the organization from legal liability and financial penalties.

- *Reducing Costs* is also a benefit. Cybersecurity frameworks can help organizations reduce the cost of security. This is because they can help organizations identify and eliminate unnecessary security controls. Those frameworks can also help increase the efficiency of response during an attack.

- Another benefit is Enhanced Business Continuity. Cybersecurity frameworks can help organizations recover from cyberattacks more quickly, as mentioned before. This is because they can help organizations develop incident response plans and test those plans according to some widely used attack methods.

In addition to these benefits, cybersecurity frameworks can also help organizations improve communication and collaboration between security teams and other stakeholders, promote a culture of security throughout the organization, and demonstrate a commitment to security to customers, partners, and regulators. Given the many benefits of cybersecurity frameworks, it is clear that all organizations should incorporate them into their information security programs, whether large or small. These frameworks can be tailored to the size and complexity of an organization.

Here are some additional tips for incorporating cybersecurity frameworks into an information security program:

1. First, start with a risk assessment. This is the first step in incorporating a cybersecurity framework. This will help you identify your organization's risks and prioritize security investments.

2. Sound governance is particularly important for a cybersecurity program. Select a framework that is right for your organization. There are many different cybersecurity frameworks available, so it is important to select one that is right for your organization's size, industry, and risk profile, as well as the complexity of your organization.

3. When it comes to implementing the framework, there are many things involved in making changes to security policies, procedures, and controls. The framework implementation, based on the size, risk, and complexity of your organization, will be especially important and critical to enhancing your cybersecurity program.

4. Once implemented, you must also monitor and continuously improve the framework, controls, and systems within your program. Monitoring it and making improvements will be done continuously and as needed. This helps ensure that your security posture is continuously moving in the right direction and proves that you are also ready for all types of attacks and all types of scenarios.

Following these tips can incorporate cybersecurity frameworks into your information security program and improve your organization's cybersecurity resiliency.

Now, we want to take a look at some well-known cybersecurity frameworks.

- NIST, which stands for the National Institute of Standards and Technology, has a cybersecurity framework, CSF. The NIST CSF is a framework developed to help organizations improve their cybersecurity posture. It is a flexible framework that can be customized to the needs of any organization, regardless of size or industry.

- Another framework is ISO/IEC 27001 and ISO/IEC 27002. These standards are developed by the International Organization for Standardization, otherwise known as ISO, and provide a comprehensive set of requirements for the information security management system, ISMS. Organizations that achieve ISO 27001 certification demonstrate that they have implemented an effective ISMS or information security management system.

- CIS controls. CIS controls are a set of 18 critical security controls that are designed to protect organizations from cyber threats. These CIS controls are based on the best practices of leading security experts and are continuously updated to reflect the latest threats. CIS controls were developed by the Center for Internet Security, CIS.

- SOC 2 reports. A SOC 2 is a type of audit or assessment that assesses an organization's security, availability, processing, integrity, confidentiality, and privacy controls over sensitive data.

SOC 2 reports are used by organizations to demonstrate their compliance with industry regulations and to reassure customers and their partners of a commitment to security. There are two types of SOC 2 reports: type 1 and type 2. Type 1 examines whether there are security controls that would meet specific objectives. Type 2 would then evaluate those controls to ensure they meet those objectives.

- PCI DSS is another framework commonly used in the financial services industries or those related to financial services. This is the payment card industry data security standard. It is a set of security requirements that organizations must meet in order to process credit and debit card payments. PCI DSS compliance is mandatory for all organizations that store, process, or transmit sensitive payment card data.

These are just a few of the many cybersecurity frameworks that are available. The best framework for an organization will depend on its specific needs and requirements.

Another step to mitigating risk is deploying advanced architectures such as zero trust architectures or zero trust network architectures (ZTA or ZTNA) and incorporating robust network segmentation. These can greatly enhance your cybersecurity program and increase your cyber resilience team. I want to spend a little time mentioning zero trust architecture; over the past few years, it has been increasingly recognized as a key component to becoming cyber-resilient. There has also been, mistakenly, no shortage of vendors claiming their product is ZTNA. Zero trust architecture is a security model that assumes no user, device, or entity, for that matter, is trusted by default. This means that all access to resources is granted on a need-to-know basis and is continuously verified.

Zero Trust Architecture, or ZTA, as it is known, is designed to protect organizations from cyberattacks, even if they have been breached.

There are many reasons why ZTA should be incorporated into the cybersecurity program. Some key reasons include improved security posture. ZTA can help organizations improve their security posture by making it more difficult for attackers to gain access to their systems and data. This is because ZTA requires users and devices to authenticate themselves and have their identities verified before they are granted access to resources. Reducing the risk of cyberattacks is also a reason to deploy ZTA. ZTA can help organizations reduce the risk of cyberattacks on separate tanks by making it more difficult for attackers to move laterally once they have gained access to a system or your infrastructure.

This is because ZTA isolates each system and only allows authorized users and devices to communicate with each other. Another reason is increased compliance with regulations. Zero Trust Architecture can help organizations comply with regulations that require them to protect their data. ZTA can help in this manner by helping organizations implement controls that meet the requirements of these regulations. Enhanced business continuity is also a reason to deploy ZTA. Enhancing business continuity makes it easier to recover from cyberattacks by deploying ZTA. This is because ZTA isolates systems and data, making it easier to contain the damage caused by a cyberattack. Another benefit is an improved user experience. ZTA can improve the user experience by making it more convenient for users to access resources due to the fact that it does not require users to constantly re-authenticate themselves. Also, reduced cost is another reason ZTA can be deployed. This can help organizations reduce the cost of security by making it easier to identify and eliminate unnecessary security controls. ZTA also helps organizations gain better visibility and control over their security posture. ZTA provides a comprehensive view of all access to resources, both internal and external.

Another benefit of deploying ZTA is improved threat detection and response. ZTA continuously monitors all access to resources and can identify suspicious activity early on, which greatly improves your response. ZTA promotes a culture of security. It requires all users and devices to be authenticated and verified before they are granted access to those resources. This promotes culture within an organization.

Another reason to deploy ZTA is that ZTA can help organizations demonstrate their commitment to security to customers, partners, and regulators because ZTA is a well-known and respected security model. ZTA has many benefits, as noted. It is clear that it should be incorporated into any cybersecurity program.

Here are some best practices for incorporating ZTA into a cybersecurity program:

1. First, start with the risk assessment. This is the first step in incorporating ZTA. It will help you identify your organization's risks and prioritize your security investments.

2. Next would be to select a ZTA framework. There are many different ZTA frameworks available, so it is important to select one that is right for your organization when it comes to size, industry, risk profile, and the complexity of your organization.

3. Once you have selected the framework, you need to implement it. This may involve making changes to your security policies, procedures, and controls. Implementing the framework that you have chosen according to the size and complexity of your

organization may be based on a risk posture that you want to achieve in the future. Not all sizes fit one company. It is important to note this and implement the framework as you see fit for your organization.

4. Once implemented, continuous monitoring and improvement are required. Companies need to monitor the framework and make improvements as they see fit. Monitoring and improving will also help you constantly evaluate attacks, techniques, policies, and procedures that may change within your organization. A security posture should consistently evolve and continuously improve as attacks change, attack methods change and an organization's business processes change.

5. Another best practice step to implement is the education of end users. This is important to educate users about ZTA and how it works. This helps them understand the need for ZTA and to comply with the new security policies and procedures. Once fully implemented, it's important to always test your ZTA implementation. This can be done on an annual basis, all the way through continuous testing. This helps identify any gaps in one's security posture and make necessary improvements. Continuously monitoring the ZTA implementation cannot be overstated. It's important to not only continuously monitor ZTA implementations but also the security controls that are working within ZTA implementations to ensure that they're working properly and protecting your organization as intended. It also helps to identify any new threats or vulnerabilities that may have emerged.

By following these tips, an organization can incorporate ZTA into its cybersecurity program and improve the organization's cybersecurity resiliency. There have been many books, articles, and white papers written on zero trust architecture, and I highly recommend reviewing some and strategizing on how you can incorporate them into your company's infrastructure or raise a topic in board and/or committee meetings.

Network segmentation is another practice to use to help mitigate risk. Network segmentation is the practice of dividing a network into smaller, more isolated segments. It may also be called micro-segmentation. This technique can help improve security by reducing the attack surface and making it more difficult for attackers to move laterally once they have gained access to a system. Network segmentation can help companies become cyber-resilient in a number of ways. Let's take a look at a few.

1. The first would be the *Reduced Attack Surface.* By dividing a network into smaller segments, network segmentation can help reduce the attack surface because each segment contains only a subset of the organization's assets, making it less likely that an attacker will be able to access all of the organization's data and systems.

2. Network segmentation also helps by *Improving Visibility and Control Over the Network.* This is because it can help identify and isolate compromised systems, which can make it easier to contain the damage caused by a cyberattack.

3. *Improve Threat Detection and Response* by helping identify suspicious traffic patterns and isolating systems under attack. Enhanced business continuity is also another way network segmentation helps organizations.

Isolating a network through network segmentation can reduce downtime for other parts of the organization. They may not even be affected by a cyberattack due to the segmentation. In addition to these benefits, network segmentation can also help improve compliance with industry regulations. For example, the payment card industry data security standard, PCI DSS, which we talked about previously, requires organizations to segment their networks to protect sensitive customer data.

Network segmentation is a valuable tool that can help companies improve their security posture and become more cyber-resilient.

Here are some best practices for implementing network segmentation:

1. The first step, as with many implementations of tools, architectures, or frameworks, is starting with the risk assessment. This helps identify assets that need to be protected and the threats that these assets face. It also helps you to start planning how you want to design your segmentation plan.

2. Once that risk assessment is done, you will design your segmentation plan. This involves defining the segments that you want to create and the type of traffic that will be allowed to flow between them.

3. The third step might involve implementing the segmentation plan. Once you've designed your segmentation plan, you will implement it. This would entail making changes to your current network infrastructure, security policies, and procedures and then evaluating that network to ensure everything is working as intended.

4. Lastly, monitor and maintain your network segmentation plan. Once you've implemented your segmentation plan, monitoring becomes very important, and maintaining it is just as important. This involves keeping the plan up-to-date as your network changes and new threats emerge. Network diagrams, including flows, protocols used, and several artifacts, need to change. Each should be updated once they are implemented in your infrastructure.

By following these tips, you can implement network segmentation to help improve your organization's security posture and continuously improve your organization's security posture.

As I mentioned, not only incorporating these advanced architectures but also actively monitoring the traffic flow across your infrastructure and managing systems in their configurations will go a long way to achieving cyber resilience.

Another way to mitigate risk from cyberattacks is by implementing advanced technologies where possible and planning for a post-quantum computing world. These are essential in a cybersecurity program. As the cybersecurity landscape is constantly evolving, with new threats emerging all the time, it is important to stay ahead of these threats. Organizations need to deploy advanced technologies in their cybersecurity programs. Let us take a quick look at some of the advanced technologies that can be deployed in a cybersecurity program.

The first one that is currently implemented across many organizations would be artificial intelligence, AI, and machine learning.

AI and machine learning (ML) are powerful tools that can be used to automate security tasks, identify threats, and respond to incidents. For example, AI can be used to analyze substantial amounts of data to identify patterns that may indicate a cyberattack. ML, or machine learning, can be used to train models that detect malicious activity.

Here's how AI can benefit a company's cybersecurity teams:

1. It provides *Increased Visibility and Control.* As I mentioned, AI can be used to analyze enormous amounts of data to identify patterns that may indicate an attack. This will help your organization gain better visibility into its security posture and identify potential threats.

2. *Improved Threat Detection and Response.* AI can be used to automate security tasks such as identifying and responding to incidents. This can help organizations reduce the time it takes to detect and respond to cyberattacks, which can minimize the damage caused by an attack.

3. *Enhanced Business Continuity.* AI can be used to help organizations recover from cyberattacks more quickly. This is done by automating the process of rebuilding systems and data and by providing insights into how to prevent future cyberattacks.

4. Another benefit is *Reduced Costs.* AI can be used to automate security tasks, which can free up security professionals to focus on more strategic initiatives. This helps an organization reduce the cost of its security program.

5. It can also help *Improve the Security Posture* by having professionals focus on more mature tasks than an organization needs to deploy. AI can be used to enhance the security posture by helping identify and mitigate advanced risks. This is done by

providing insight into the security posture of an organization and by automating security tasks associated with the organization. This helps an organization become more resilient. But the more automation an organization can use, the more mature they become, and the more cyber-resilient that organization should then become.

6. AI also improves the *User Experience*. This is done by making it easier for your security professionals to use the security tools and by providing more personalized security recommendations for each organization.

7. AI also *Helps Organizations Comply With Regulations* requiring them to protect their data through automated processes to help it comply with security controls. AI can help identify controls that are not in line with what the organization intended and then make recommendations on how to enhance those security controls.

AI is a powerful tool that can be used to improve the security posture of an organization. AI provides better visibility and control over the security environment, helps improve threat detection and response, enhances business continuity, reduces costs, and improves cybersecurity resilience.

Here's some additional information on deploying AI in a cybersecurity program. As mentioned above, always start with a risk assessment. This helps identify areas where you want AI to enhance your program.

The next step might be selecting the right AI tools. Many different AI tools are available, so it's important to select the right tool for your organization. Then, you want to train your AI models.

Once you've selected AI tools, you need to train them on your organization's data. This can take a little bit of time, or it can be done over weeks. It is also a continuous process to train your models.

AI tools learn to identify and respond to threats that are specific to one's organization. The more data you can feed to an AI model, the better that model becomes. More data occurs over a longer period of time.

Once these AI models are trained, it's important to monitor and continuously evaluate them. This will help an organization and the AI models they have deployed perform as expected and be able to identify and respond to new threats.

Many AI tools exist to help companies increase the efficacy of their cybersecurity programs. I suggest looking at what is available, deploying some available tools, and then reevaluating to see if your organization needs to develop its own AI models. Developing may require a lot of time and resources from within your organization.

By following these tips, organizations can use AI to improve their cybersecurity posture and resilience.

Blockchain is another advanced technology that can be used to enhance your cybersecurity program. Blockchain is a distributed ledger technology that can be used to store and track data in a secure, immutable manner. This makes it ideal for storing sensitive data, such as financial information or medical records.

Blockchain can also be used to track the provenance of goods and services, which can help prevent counterfeiting and fraud. This protects your intellectual property.

Let's discuss some ways blockchain can benefit a cybersecurity program.

1. It provides *Increased Visibility and Control.* Blockchain can be used to create a single, immutable record of all transactions that have ever taken place on a network. This helps an organization gain better visibility into its security posture and identify potential threats.

2. Another benefit is *Improved Threat Detection and Response.* Blockchain can be used to track the provenance of data, which can help organizations identify suspicious activity. An example might be if a piece of data is found to be fraudulent. Blockchain can be used to trace its origin and identify the source of fraud.

3. *Enhanced Business Continuity.* Blockchain can be used to store critical data securely and tamper-proof, as mentioned. This will help an organization recover from a cyberattack more quickly.

4. *Reducing Costs* is also a benefit of blockchain. Blockchain can be used to automate security tasks, which in turn frees up security professionals to focus on other tasks and helps reduce the cost of a security program.

5. Blockchain can be used to *Improve the Security Posture* of an organization. By providing insights into the security posture of an organization and automating those security tasks, organizations can mitigate risk more quickly.

6. Blockchain provides an *Improved User Experience* by making it easier for users to use security tools and by providing more personalized recommendations around the use of those tools. Here are examples of how blockchain is being used in

cybersecurity programs currently. Blockchain is used to track the movement of goods and materials in a supply chain. This helps prevent counterfeiting and fraud, as mentioned above, and it ensures that products meet quality standards. This protects an organization's intellectual property.

7. Blockchain is also used in *Identity Management*. It can be used to create a secure and tamper-proof record of identities. This helps protect against identity theft and fraud. In payments, blockchain is used to make secure and anonymous payments. This helps reduce the risk of fraud and cyberattacks.

In healthcare organizations, blockchain is used to store and track patient records in a secure and tamper-proof manner. Again, this helps improve patient care and reduces the risk of data breaches. In financial services, blockchain is used to record and track financial transactions in a secure and tamper-proof manner. This prevents fraud and helps improve the efficiency of financial services.

Blockchain is a promising new technology that has the potential to revolutionize cybersecurity. I still believe many untapped uses for blockchain in the cybersecurity program are ripe for development. By leveraging blockchain, organizations can gain better visibility and control over their security environment, improve their security, improve threat detection and response, enhance business continuity, reduce costs, and improve cyber resilience.

Here are some ways and best practices to deploy blockchain in cybersecurity:

1. Again, starting with the *Risk Assessment*. I know I say this over and over, but that is usually step one. This helps identify assets that need to be protected and threats that these assets face.

2. Selecting the right blockchain *Platform* is second. This is done by following proper governance. There are many different types of governance models to use, as well as many different types of blockchain platforms. It's important to select the right platform that fits your organization's name.

3. *Implement the solution.* An organization can do this on their own, or they can utilize expertise from other companies, consulting firms, and the like to deploy the blockchain platform with their team.

4. Once deployed, *Monitoring and Continuously Evaluating* would be the next step. This helps ensure that the blockchain solution you have deployed is performing as expected and that you are able to protect your organization's assets from threats when deploying blockchain technologies.

A third advanced technology is quantum computing. Quantum computing is a new computing paradigm that has the potential to revolutionize many industries, including cybersecurity. Quantum computers are much faster than traditional computers and can be used to break many of the encryption algorithms that are currently used to protect data. They can also be used to protect against the breaking of encryption algorithms.

Here are some ways that quantum computing can help increase the efficacy of a cybersecurity program at your organization.

1. Quantum computing helps *Improve Threat Detection and Response* because it can be used to analyze large amounts of data much faster than traditional computing. This helps organizations identify and manage responses to threats more quickly.

2. Quantum computing *Enhances Business Continuity* as well. This is done by the speed with which quantum computing can help rebuild systems and data more quickly after a cyberattack. This helps organizations recover from a cyberattack more quickly.

3. Quantum computing also *Reduces Costs* by automating many of the tasks that are currently performed by security professionals. This helps reduce the cost of a cybersecurity program and its resources. It also frees your organization and the security professionals to do more advanced strategic initiatives.

4. Quantum computing *Enhances the Security Posture* of an organization. Quantum computing is used to develop new security tools and techniques that are more effective at protecting data from cyberattacks.

Here are some additional examples of how quantum computing is currently being used in cybersecurity:

- Quantum computing-resistant cryptography. Quantum computers can be used to break many of the encryption algorithms that are currently used to protect data.

- Quantum-resistant cryptography is a new type of cryptography that is designed to be secure against quantum computers. Organizations can deploy quantum-resistant cryptography to protect against these types of attacks.

- Quantum machine learning. Quantum machine learning is a new type of machine learning that can be used to analyze large amounts of data more quickly and accurately than traditional machine learning. Organizations can deploy these techniques to quickly learn, identify, and respond to threats.

- Quantum network security. Quantum networks are more secure than traditional networks because they can use quantum cryptography to protect their data. Organizations can use quantum networks to transmit sensitive data more securely.

Quantum computing is a promising new technology that has the potential to revolutionize cybersecurity.

Here are some ways to deploy quantum computing in a cybersecurity program:

1. As always, we start with a *Risk Assessment*. This helps you understand how quantum computers work and how they could be used to attack your organization.

2. Once that is done, you will *Identify the Assets* you want to protect. Understand these threats. Once you understand the threats, you must identify the assets that need to be protected from quantum computers or utilized to protect your organization.

3. You would also then *Select the Right Quantum Computing Tools*. Selecting the right tool for your organization to meet the use cases you have designed to undertake is important.

4. You would then *Implement These Tools and Continuously Monitor and Evaluate* these solutions to ensure they're helping your organization how you have designed them.

Just as quantum computing can be used to help your organization, there are many ways that threat actors are now starting to use quantum computing to hurt your organization. Some of those include breaking current encryption algorithms. As threat actors start deploying quantum computers, they're able to break traditional cryptography much quicker than they would in the past. This allows them to steal sensitive information such as personally identifiable information, credit card numbers, medical records, and passwords.

Quantum computers are being used to develop new cyberattacks and new cyberattack techniques. Quantum computing deployed along with AI and machine learning can quickly learn an organization's protection techniques and then design attack methods that would break those techniques. This makes it more difficult for an organization to protect itself against cyberattacks.

In a post-quantum computing world, it's important to prepare for the impact of quantum computing on a cybersecurity program.

Here are some steps organizations can take:

1. Start by understanding the threat. What is quantum computing? What are the threats my organization faces with quantum computing? This includes how quantum computers work and how they could be used to attack your organization and, in turn, deploying techniques that could help prevent that.

2. Have a good understanding of your current cybersecurity posture. Once you understand these threads, understanding your cybersecurity posture and identifying assets that are most at risk and the gaps in our security controls can help us defend against quantum computing attacks.

3. Implementing quantum-resistant cryptography where needed is also a new technique to secure against quantum computers. Organizations should implement quantum-resistant cryptography for any data that needs to be protected from

quantum computing, such as sensitive data, personally identifiable information, data related to intellectual property, and many more.

4. Organizations should invest in quantum security research. Quantum security research is still very much in its early stages, but staying up-to-date on the latest developments is very important. This will help organizations understand the latest threats and develop new security controls to protect their organizations. This can also help an organization build quantum computing into its security program before the threat actors have a chance to develop quantum attack methods to attack the organization.

By taking these steps, an organization can help prepare for the impact of quantum computing on cybersecurity.

These are just a few of the advanced technologies that can be deployed in a cybersecurity program. Others would include micro-segmentation and zero-trust architecture, as we've talked about intrusion prevention systems, data loss prevention, which is a security concept, and solutions that help organizations protect sensitive data from being lost or stolen.

Next-generation endpoint security, across server workstations IoT, Internet of Things devices, and web application firewalls, is a system that protects web applications from cyberattacks.

Also, ICS (industrial control systems) control security. Restating ICS industrial control system security models. By deploying these technologies, organizations can improve their security posture and protect themselves from very sophisticated cyber-security attacks.

In addition to these technologies, let's not forget about the fundamentals that we talked about previously. If you cannot get the fundamentals right, deploying advanced technologies will only hinder your organization and your cybersecurity program. Fundamental best practices include keeping software up-to-date, vulnerability management, patch management, implementing strong passwords and multifactor authentication, or even extending to no-password technology.

At a minimum, you should deploy multi-factor authentication. Education, educating employees about cybersecurity and cybersecurity threats, conducting regular security assessments, and having a plan for responding to cyberattacks are all fundamental to a program. By following these best practices, organizations can further improve their cybersecurity posture and protect themselves from cybersecurity attacks.

Another step to mitigate risk is incorporating cyber threat intelligence and cyber threat hunting into your organization. Cyber threat intelligence includes collecting and analyzing data to identify and mitigate threats. Both cyber threat intelligence and threat hunting are important components of a comprehensive cybersecurity program.

Cyber threat intelligence, or CTI, can come from a variety of resources, such as publicly available resources, including news articles, blog posts, and other publicly available information. It can also come from private sources. This includes information from security vendors, threat intelligence providers, government organizations, and other organizations that collect and share threat intelligence. It can also come from internal resources, including security logs, incident reports, and other data within your organization. CTI can be used to identify new threats, understand the tactics, techniques, and procedures of threat actors, and then help you prioritize your security efforts.

Threat hunting is the process of actively searching for threats in your infrastructure that may have evaded your existing security controls. This is typically done by analyzing your security logs and other data for indicators of compromise, otherwise known as IOCs. IOCs are specific pieces of data that can indicate the presence of a threat, such as a specific IP address, a file hash, or a network pattern.

Threat hunting can help an organization identify threats that its existing security controls may not have detected.

Here are some helpful ways to incorporate CTI and threat hunting into a cybersecurity program:

1. First, you'd want to *Define Your Goals* for the program. What do you hope to achieve by incorporating CTI and threat hunting into your program? Do you want to identify new threats, understand threat actors' tactics, techniques, and procedures, or prioritize your security efforts? Maybe it's all of them.

2. Then, you'd want to *Gather Data*. You need to gather data from a variety of sources in order to generate effective CTI and threat-hunting insights. This data can include publicly available information, private information, information from your organization, as well as many other sources.

3. *Analyzing the Data* is also very important and would be the next step. Once this data has been gathered, it needs to be analyzed. This can be done through manual means, automated tools, or a combination of both.

4. Once you have analyzed this data and realized what is important to your organization, *Disseminating the Data*, the analysis, and the insights to the appropriate people in your organization is

important. This could include two security analysts, incident response teams, your IT teams, business leaders, or other stakeholders. Even sharing this data with external partners or the community as a whole, once it's been scrubbed, is also important.

5. After you've done these things, you want to *Continuously Monitor and Evaluate* your program, the CTI you're receiving, and the threat-hunting techniques you're undertaking. This may include tracking the number of threats you're identifying, the accuracy of the insights, and the program's impact on your overall cybersecurity, resilience, and posture.

Some organizations that regularly disseminate cyber threat intelligence include a program by the FBI called InfraGard. It may also include information sharing and analysis centers or information sharing and analysis organizations, otherwise known as ISOC and ISOC forward slash ISAO. Some cybersecurity companies that you may deploy throughout your organization and from your peers are good ways to garner threat intelligence for your organization.

By following these best practices, you can incorporate CTI and threat hunting into your cybersecurity program and enhance your ability to detect and respond to threats.

Another important step in mitigating cyber risk is incident response. Incident response is the process of responding to and recovering from a cybersecurity incident. This includes automation and orchestration and the importance to your organization of effective communication, who you're going to communicate to, whether internal, external, or both, and then the collaboration during a security incident.

Incident response includes a variety of steps:

1. Generally, *Detecting the Incident* is the first step. This can be done by monitoring for suspicious activity such as unusual logins, unauthorized access to systems, system behavior in a different manner, or anomalous behavior.

2. You would then *Investigate the Incident* to determine the full extent of the damage.

3. Then, you'd want to *Contain the Incident* after investigating it and understanding what it was doing. You may want to isolate affected systems, remove that malware, reset passwords, or do a combination of many things. If there are any systems that are damaged, restoring those systems and the related data would be the next step.

4. Then, as a final step, you might have a *Lesson-Learned Meeting* within your organization. This may help you develop additional procedures or practices to improve incident response in the future.

A well-defined incident response plan is essential for any organization.

Here are some of the benefits of having a strong incident response plan and program:

1. *Reduce Financial Losses.* Financial losses can quickly mount from a data breach. A robust, well-tested incident response plan can help you reduce those financial losses. It also helps an organization protect its reputation. Many companies have to report that they have been breached. In fact, new laws for public

companies are underway as I write this, that will require all public companies to report material breaches.

2. *Compliance with Regulations* is also a benefit of a robust incident response plan. By having a plan in place, businesses can demonstrate to regulators that they comply with current regulations and can respond to an incident effectively.

3. Having a strong incident response plan also greatly enhances your cybersecurity *Posture and Resilience*. These plans can help identify gaps and then help you improve those security gaps.

What are some incident response frameworks that companies can deploy? These frameworks, which are a step-by-step guide that an organization can use to respond to a cyberattack, provide a systematic approach to incident response and can help an organization minimize the impact of a cyberattack.

Here are a few of the more popular incident response frameworks:

- *The NIST Cybersecurity Framework* (CSF) is a risk-based framework that can be used to improve an organization's cybersecurity posture. The Cybersecurity Framework includes a section on incident response that provides guidance on preparing for, detecting, responding to, and recovering from cyberattacks.

- Another framework is the *MITRE ATT&CK* framework or ATT&CK knowledge base. The MITRE ATT&CK framework is a knowledge base of adversary tactics, techniques, and procedures (TTPs) based on real-world observations. The attack framework can be used to understand the threat landscape and develop an incident response plan tailored to the specific threats your organization may face.

- Another framework is known as the *Computer Security Incident Response Team (CSIRT)* handbook. It is a guide to establishing and operating incident response planning, incident handling, and post-incident activities.

- *The National Institute of Standards and Technology (NIST) Cybersecurity Framework (CSF)* provides a set of recommendations for organizations to improve their cybersecurity posture.

The Cybersecurity Framework, CSF, is divided into five functions:

1. *Identify*: identify the assets that need to be protected and the threats that those assets face.
2. *Protect*: implementing security controls to reduce the likelihood of a cyberattack.
3. *Detect*: identifies and responds to potential or actual threats.
4. *Response*: responding to a cyberattack to minimize its impact
5. *Recovery*: recovering from a cyberattack and resuming normal operations.

The CSF incident response function includes the following steps:

1. *Prepare*: develop and maintain the incident response plan.
2. *Detection:* we mentioned above is identifying and investigating potential or actual threats.
3. *Containment*: isolate the affected systems and data to prevent further damage.
4. *Eliminate*: removing the threat and restoring systems to data and to normal operations.
5. *Recover*: restore normal operations and learn from the incident.

The NIST CSF is a flexible framework that can be customized to meet the specific needs of any organization.

Here are some ways to implement the NIST CSF's incident response function.

1. *Assign Roles and Responsibilities.* Assigning specific roles and responsibilities for incident response to different individuals or teams helps ensure that all aspects of incident response are covered.

2. *Create an Incident Response Plan* is another step to deploying. This should be a detailed document that outlines the steps that will be taken in response to a cyberattack.

3. The plan should be *Tailored to the Organization's Specific Needs* and updated regularly. In addition to the plan, you may have supplemental material that goes through each type of attack vector that a threat actor could use to harm an organization, such as ransomware, DDoS attacks, insider threads, general mail malware, restate, general run of the mail malware, and many other types of attacks.

   An organization could have multiple playbooks, including playbooks on industrial control systems versus the corporate network, for example.

4. The next step would be *Testing the Plan.* The incident response plan should be tested regularly. This is recommended to be done annually, quarterly, or more often. An organization can incorporate fire drills, which are small, specifically tailored incident response exercises. Testing of the incident response plan

can be done using simulated cyberattacks or real-world data from previous incidents, or you can do tabletop exercises in mock scenarios. All incidents should be documented in detail, with the documentation then being used to improve the organization's incident response plan and to track the effectiveness of the plan over time.

5. The NIST CSF incident response function includes *Training Employees*. All employees should be trained on the organization's incident response plan and specific steps to take if they suspect a cyberattack. This may involve general user training, but it may also involve training important stakeholders that are part of the incident response plan on what they might need to do, such as your IT department, definitely your cybersecurity department, legal, finance, and communications departments, as well as any partners that you might use known as Manage Security Providers or Manage Security Service Providers or MSSPSs. You would also continuously monitor and improve this incident response plan. This helps ensure that the plan is always up-to-date and that it is effective in responding to cyberattacks.

Now, let's look at selecting an incident response framework that we may have discussed above. There are many others besides the three that we talked about.

When selecting an incident response framework, organizations should consider the following factors:

1. One is the *Size and Complexity of Their Organization*. Large organizations may need a comprehensive incident response framework compared to smaller ones.

2. Organizations should also *Consider the Industry in Which They Operate*. Different industries face different types of cyberattacks.

3. Organizations should select a framework that is *Tailored to Their Specific Threats*. They may even select multiple frameworks based on their organization's risk profile and develop their own framework.

4. *The Organization's Budget* should also be factored in. These frameworks and their implementation vary in price. An organization should select an incident response framework that fits within their budget, and then they should also tailor that framework to their organization.

5. *Customization* of an incident response framework is very common. Once an organization has selected the framework, they should customize it to fit their specific needs. It may be enacting the entire framework or just pieces of the framework. You may have to remove steps or include additional steps that are not in the framework that you've selected. It's important to modify a framework to fit your organization's needs, incorporating specific processes and procedures that are tailored to your organization.

By following these steps, organizations can select, customize, and implement an incident response framework that would help them minimize the impact of a cyberattack.

In conclusion, incident response is an essential part of any cybersecurity program. By having a plan in place and regularly testing and updating the plan, businesses can be prepared to quickly identify and respond to incidents, minimizing damage and disruption.

Next, let's take a look at monitoring third- and fourth-party security postures to minimize the risk to your organization.

As we know, in today's digital world, businesses of all sizes rely on third-party vendors, and those vendors may have fourth parties. These vendors provide a wide range of services. These vendors can include cloud providers, software developers, and IT support providers. They can also include operational-type vendors, such as those that maintain your HVAC systems, or they may outsource functions in accounting and business operations of all kinds. Third- and fourth-party vendors provide a wide range of services which are critical to a corporation.

While third-party relationships may be very beneficial to businesses, they also introduce new cyber-security risks. Third-party vendors may have access to sensitive data such as customer PII (personally identifiable information), financial information, and intellectual property. They may also have the ability to manipulate or disrupt critical business systems. If a third-party vendor is compromised or one of their fourth parties,

Third-party vendors may have access to sensitive data such as customer PII, financial information, and intellectual property. They may also have the ability to manipulate or disrupt critical business systems. If a third-party vendor is compromised, it can have a devastating impact on the business that that vendor has partnered with. It is critical that we state it. That's why third-party risk management is so critical. Third-party risk management is the process of identifying, assessing, and mitigating the risks associated with third-party relationships.

A comprehensive third-party risk management program might include the following steps:

1. Identifying all third-party vendors, both direct and indirect.

2. Assessing the risk of each of these vendors, which would include evaluating the vendor's security posture, financial stability, and compliance with applicable laws and regulations.

3. Implementing controls to mitigate any residual risk. This might require vendors to sign confidentiality agreements, implement additional security controls, or undergo regular security assessments.

4. Monitor the effect of the controls in this program. This would involve conducting regular audits and reviewing these vendors to ensure the controls are still effective.

Third-party risk management is an ongoing process that should be continuously reviewed and updated. By implementing a comprehensive third-party risk management program, businesses can help protect themselves from the cybersecurity risk associated with third-party relationships.

Some benefits of implementing a third-party risk management program might include:

1. *Compliance with Regulations.* Many industries are subject to regulations that require them to manage third-party risk.

2. Another benefit is the *Reduced Risk of Data Breaches.* This helps identify and mitigate the risk associated with data breaches from a vendor. This helps protect your organization from the financial and reputational damage that can result from a third-party data breach.

3. It can also help you *Mitigate the Risk* by looking at the controls within your organization that are impacted by this third-party vendor.

Third-party risk management helps improve operational efficiency by reducing the number of possible security incidents. This frees up IT resources to focus on other priorities.

Conclusion: third-party risk management is an essential component of any cybersecurity program. By implementing a comprehensive third-party risk management program, businesses can help protect themselves from cybersecurity risks associated with third-party relationships.

Other ways to mitigate risk might include deploying attack simulations and strong user awareness programs to enhance your cyber resilience. It may also involve collaborative efforts with your peers, other industries, and many other organizations to improve the robustness of your cybersecurity program. These are just some of the many ways that your company can mitigate the risk of cybersecurity attacks.

It's important to note that no one size fits all. There are many organizations across the globe that have to comply with many different regulations and standards. Small organizations, which many may think are not susceptible to cybersecurity attacks, are, in fact, threat actors. Consider them a soft target, so they should also deploy robust cybersecurity principles according to their risk profile. Large organizations are constantly under attack. They need to deploy the most sophisticated risk mitigation steps so they can meet the challenges of current cyber threats as well as evolving threats.

To meet the cybersecurity risk that we all face in today's digital world with growing attack vectors, it's important to note that we are all part of this ecosystem. The more we can collaborate with one another and share information across boundaries, the better we'll all be at defeating threat actors who try to attack us and harm our critical infrastructure.

It's important to utilize the resources that we all have at our disposal and give back to the community in many ways. Mitigating risk is critical to companies defeating cybersecurity attacks. There are many risk mitigation efforts, as we've discussed. An organization should start small and then progress to a larger risk mitigation program. It's important that all people in the organization understand the risk you're trying to mitigate. These principles will help an organization become very cyber resilient, meet today's cyber threats, and succeed when they are attacked.

From the Desk of a CISO: A CISO's Perspective

Cybersecurity Culture Featuring Mario Chiock

Mario Chiock Bio: Mario Chiock has over 40 years of experience in oil field operations, IT, cyber security, risk, compliance, privacy, and auditing. Before becoming Schlumberger (SLB) Fellow Emeritus and CISO, he was responsible for developing the company's worldwide, long-term cyber security strategy, including its digital transformation and cloud migration. He is recognized for his leadership and expertise in all aspects of cybersecurity, both within the company and in the broader community. In the oil and gas sector, he is considered a subject matter expert in cloud technologies, industrial internet of things, mergers/acquisitions, and digital transformation.

Mario has been an active member of the Information Systems Security Association (ISSA) for over 20 years. He has held numerous board positions in the Austin Capital of Texas Chapter, as well as the South Texas Chapter in Houston. He served as the president of the South Texas Chapter in 2007 and brought home the "Chapter of the Year" award. He continues to serve on the board. Mario also actively volunteers as a trainer for security certifications such as CISM, CISA, and CRISC and has mentored many successful CSOs and CISOs in the Austin and Houston areas. He is also very active with Evanta, serving as a speaker and instructor for their CISO Institute. In 2015, he spoke at the SPE and API conferences, and in 2016, he was a panelist at the GEO2016 conference in Bahrain. In 2018, he was awarded the South Central Region InfraGard Award for "INMA Leadership."

Mario was recognized as one of the top 25 security executives out of more than 10,000 in the ExecRank 2013 Security Executive Rankings. He also won the 2012 Central Information Security Executive (ISE) "People's Choice Award." In 2014, he received the CSO40 – 2014 award, was named an "ISSA Fellow," and won both the ISC2 Americas Information Security Leadership Awards (ISLA) and the "ISSA Honor Roll" award. In 2017, he received the InfraGard Houston Award of Excellence for the Private-Public Partnership in Cybersecurity.

He is an active member of the Houston security community, providing security talks, training, and volunteering his IT security expertise to local non-profit organizations. Currently, Mario is a board member of the Houston InfraGard Chapter. He has also served on the executive and technical advisory boards of numerous security companies, including WatchFire (acquired by IBM), ISS (acquired by IBM), and Demisto SOAR (acquired by Palo Alto Networks). Currently, he is active on the executive advisory boards for Palo Alto Networks and Qualys and serves on the Google Cloud Platform advisory board. He is also a strategic advisor to Onapsis and a past board member for various organizations.

Mario holds CISSP, CISM, and CISA certifications. He has previously served as the chair of the American Petroleum Institute Information (API)

Security Sub-Committee and was instrumental in the formation of the Oil & Gas ISAC (ONG-ISAC).

> *"Strength lies in differences, not in similarities."*
> – Stephen R. Covey, educator, author, and businessman

> *"No man will make a great leader who wants to do it all himself, or to get all the credit for doing it."*
> – Andrew Carnegie, industrialist and philanthropist

Driving a Culture of Cybersecurity to Enhance Corporate Cyber Resilience

Organizational culture is a reflection of values, behaviors, and shared beliefs within that organization. Companies must drive a culture of cybersecurity and extend cybersecurity teams beyond traditional technology roles. By instilling a culture of cybersecurity, we are emphasizing that the responsibility of safeguarding our organization's data and intellectual property is not just confined to the security or IT department; it belongs to the entire organization, and the tone at the top is particularly important.

Culture change or creation is not an overnight task or quickly achieved. A culture inclusive of cybersecurity is a journey that requires continuous actions throughout the year that involve cybersecurity in onboarding, awareness, education, rewarding employees, and open communication.

A CISO must collaborate with all departments in an organization to help drive change.

"My role in culture change as the CISO is to work with the HR department to collaborate with HR to be responsible for the development and awareness training for every employee in the company," - Mario Chiock

A company should recognize that to be a secure organization, technology is not the most critical aspect of your program; it is people who are your most important asset when it comes to building an organization that has a high level of security awareness. The HR department is key to helping drive employee initiatives, and CISOs must utilize their expertise in the development of people in an organization.

"I found that conducting tabletop exercises and crisis management drills at the executive level actually helps drive culture change... It is a learning experience to help them learn that it is about people and not only technology and processes" - Mario Chiock

When it comes to building teams, all companies need an extension of their cybersecurity team. The concern of cyber hygiene should not just lie within the cyber and/or IT teams; it is an organizational concern and should extend throughout the organization and external partners.

Human resources can play a pivotal role by incorporating cybersecurity into job descriptions, performance reviews, and training programs. They can ensure that cybersecurity is embedded in the recruitment, retention, and growth processes of employees. CISOs should engage non-IT departments, whether the executive leadership team, finance, communications, operations, legal, or all others. Engaging them in cybersecurity initiatives ensures that they follow best practices relevant to their specific roles.

Creating cybersecurity champions in various departments who act as the primary point of contact for their teams can assist in driving culture within an organization and assist in the 'marketing' of cybersecurity.

"Cybersecurity champions were rewarded and made part of the cybersecurity team, an extension of our team. I built in a scholarship-type program where I included in the budget funding for them to go to conferences that their manager may not send them to" - Mario Chiock

Consider bringing in external cybersecurity consultants or firms in areas where in-house expertise may be lacking. They can provide a fresh perspective, identify blind spots, and introduce globally accepted best practices.

"I consider cloud service providers cybersecurity champions when we go through the negotiation of contracts. We ensure they are part of the extended cybersecurity team. This is why third-party management is very important" - Mario Chiock

Effective communication is critical to achieving a mature cybersecurity program. In the discipline of cybersecurity, where seconds can matter, fast response and coordinated actions are essential to mitigate cyber threats and the actions of threat groups. Communication serves as the catalyst that brings together all facets of the program. Effective, well-intentioned communication ensures that security teams are well informed about the latest threats and vulnerabilities, enabling them to proactively defend against emerging risks.

Moreover, communication fosters collaboration among various stakeholders, from IT professionals and security experts to management and end-users, creating a unified front against cyber threats. Clear and timely communication also plays a critical role in incident response, allowing organizations to contain and mitigate breaches swiftly while minimizing damage. Communication is the linchpin that transforms a collection of people, security measures and processes, and tools into an agile cybersecurity program.

"We bring our board of directors to our Security Operations Center once a year to show them how we operate and communicate our program during that time. My first recommendation to incoming CISOs who are not experienced in presenting to the Board is to find a mentor, which could be someone else in the industry, in the company, or others, that will provide you with advice on presenting to the Board" - Mario Chiock

A robust cybersecurity posture is not just about having the latest technology and tools; it's about people and processes. Driving a culture of cybersecurity means embedding security throughout the entire organization, where every individual understands, values, and acts on it. When building extended cybersecurity teams, we acknowledge that in the fight against cyber threats, every employee, irrespective of their role, is a valuable ally.

Communication is what binds a company's cybersecurity program together. CISOs, in their complex roles, should understand that communication styles must change based on the audience they are speaking to. As stated in a recent CSO magazine article, "The last decade or so has been a hard-fought battle for CISOs in simply gaining the ear of their boards of directors, let alone joining their ranks." The effort CISOs and cybersecurity teams take to build a company's cybersecurity program should not be undermined by a lack of communication.

CHAPTER 6

The Board of Directors and CISO Collaboration: Mature the Program

It has been years since cybersecurity executives have been talking about communicating between themselves and the board of directors. The evolution of this conversation has gone from properly asking for a budget to making a business case for a budget to reducing and/or mitigating cyber risk to becoming a cyber-resilient organization. It is now quite common at cybersecurity conferences that communication with the board will be one of the top three concerns CISOs raise. In a recent article by JM Search titled Six Cybersecurity Trends Influencing CISOs Today, published in the fall of 2021, effective board and executive team communications were listed as one of the top concerns. There are numerous books, articles, webinars, presentations, eBooks, and the like written on this topic.

Compounding this challenge and increasing anxiety is the fact that many CISOs either lack experience in public speaking—such as keynotes, panels, webinars, and other similar engagements—or they lack training in areas like executive speaking, Toastmasters, or other preparatory exercises that would help them feel moderately comfortable addressing the board.

Additionally, despite all the security tools that exist, we in the cybersecurity community lack a comprehensive tool to measure the ROI of a security program or demonstrate its business value. Sure, it's possible to demonstrate certain aspects, such as the cost per record of a breach, qualifying, conducting more regular phishing training over an extended period of time, or utilizing an MSSP Managed Security Service provider for SOC (Security Operations Center) services. However, tools used in the business ecosystem, like Salesforce Workday, Google Analytics HubSpot, Domo Server, CRM, or similar, are hard to come by in cybersecurity.

With the SEC's new rule on cybersecurity risk management strategy, governance, and incident disclosure enacted in July 2023, along with other regulatory entities releasing new and/or additional cybersecurity regulations and mandates, the relationship between the CISO and board becomes even more important. Cultivating the relationship and deepening it is not an easy task. It requires time to build trust, time to understand the message that should be delivered for effective results, and time to grasp the cyber risks that companies face. Underlying each of these complexities is the statistic that the average tenure of a CISO is estimated to be 18 to 26 months, about half of that of a CIO, even less when compared to that of a CEO.

In the "2023 Director's Handbook on Cyber Risk Oversight" published by the National Association of Corporate Directors (NACD), Jen Easterly, Director of the Cybersecurity and Infrastructure Security Agency, wrote in her foreword titled "Sustainable Cybersecurity: Thinking Bigger in Our Approach to Resilient Infrastructure and Customer Safety" that we need a new model of sustainable cybersecurity. This model starts with a commitment at the board level to incentivize a culture of corporate cyber responsibility. In this culture, managing cyber risk is treated as a fundamental matter of both good governance and good corporate citizenship.

The collaboration between the board of directors and a Chief Information Security Officer is vital in the corporate world due to today's cyber risk climate. Strong collaboration helps to ensure that the company is adequately protected from cyber threats and that CISOs are able to align their program to the board's and company's risk tolerance. The board of directors is responsible for governing a company's overall operations, including cybersecurity, and the CISO is responsible for the security of the company's information systems and data.

The board of directors and CISO can better identify and assess cyber risk by working together. Boards can ensure CISOs are provided with the resources they need to identify and assess cyber risk. The CISO can, in turn, provide the board with regular updates on the company's cyber risk program and the organization's cyber posture. Boards may also ensure that any and all cyber policies are implemented, up-to-date, and enforced. The board can monitor any road maps and strategies a CISO might develop while working with the CISO and executive team to ensure the company is making adequate process progress.

Just as important, if not more so, when a cyber incident occurs, the board of directors will expect to be informed during the response related to the incident and work as a team, along with others in the organization, to mitigate the incident.

When a company views cyber risk not just as the security team's issue to mitigate but as a company-wide risk issue, collaboration and involvement will integrate cyber risk into the company's culture. Collaboration across the hierarchy will occur from the board down, from executives to their direct reports, and across all levels of the organization.

The benefits of strong collaboration between the board of directors, Executive Management, and the CISO can include the following: There is an improved cybersecurity posture. By taking steps to improve cybersecurity, the board of directors, the CISO, and other company executives can help reduce the risk of data breaches. A company that is seen as being forward-thinking in cybersecurity is likely to have a better reputation with customers, investors, and other stakeholders. By working with the CISO, the board of directors can help ensure that the company complies with applicable cybersecurity regulations, whether new rules by the SEC or notification laws required by states and countries.

Strong collaboration between the board, company executives, serving on applicable board committees, and the CISO can also improve a company's cyber resilience—a company's ability to anticipate, withstand, recover from, and adapt to adverse conditions, attacks, and/or compromises on systems that utilize cyber resources in a diverse number of ways.

These include increased visibility of cyber risk. The board of directors is responsible for overseeing the company's overall risk management, including cyber risk. However, the board may not have the technical expertise to fully understand the company's cyber risk. By collaborating intently with the CISO, the board can better understand the company's cyber risks and how to mitigate them. Improved decision-making.

The board of directors is additionally responsible for making decisions about the company's cybersecurity posture. However, these decisions can be difficult to make without clear and accurate information about the cyber risks a company is facing. It takes effective communication with the CISO, and then the board can make more informed decisions about the company's cybersecurity posture and the tasks that need to be undertaken to reduce risks and improve the program holistically. CISOs do not get a pass at this either; it is a two-way street that takes considerable involvement from both parties.

The board of directors must trust the CISO to provide them with accurate and timely information about the company's cybersecurity posture. By communicating effectively with the board, the CISO can help build trust and confidence between the two parties.

If a cyber incident occurs, the board of directors must be able to respond quickly and effectively. By including the board in incident response plans, which include communication from the CISO related to an event escalating to the board according to incident level, the board can gain a better understanding of the incident, how the company is responding to it, and the overall status of mitigating the incident.

Effective communication between the board of directors and CISO may include the following best practices: The CISO should provide regular updates to the board on the company's cybersecurity posture and approach to cyber risk. These updates should be clear, concise, and written in a communication style that the board will understand. It is important to understand that technical jargon, more than likely, will not be welcomed by the board. Speak the board's language.

Another best practice is that a CISO should be willing to answer the board's questions about cybersecurity and help the board fully understand the company's cyber risk. The CISO must be prepared to answer questions in a way that the board can understand.

The board should also be willing to take the advice of the CISO. The CISO should be an expert on cybersecurity; hence, the board should be willing to listen to their recommendations and take them into account when making decisions about the company's cybersecurity initiatives and the overall posture of the company.

The board and the CISO should work together, along with other executives, to develop a cybersecurity culture within the company. This culture should emphasize the importance of cybersecurity and the need for all employees to be aware of the company's cyber risk.

CISO Responsibilities

The board of directors is now more concerned with cyber risk than at any time previously. With new regulatory requirements, an escalating risk environment, and business digital transformation, boards must track the pulse of cybersecurity. In order for the CISO and the board to work effectively together, it is important for them to have a strong working relationship. This means that they need to communicate regularly, share information, and work together to develop and implement cybersecurity policies and procedures. CISOs, in turn, need to keep identifying risks and translating those risks into communication once the board can understand them so that they may incorporate cyber resilience into the company's strategic planning.

Several practices can help the CISO achieve strong results. The CISO should have a direct line of communication with the board and be allowed to provide regular updates on cybersecurity risks and threats in person, through a committee, or through regular reporting. The point is to communicate often.

Another practice is that the CISO should meet with the board on a regular basis, and that meeting could be held quarterly with discussions related to cybersecurity risk, the company's cybersecurity posture, strategic roadmap goals for the cybersecurity program, along with updates and any other topics concerning cybersecurity that your board may be interested in. These meetings can occur quarterly, annually, or at any other time acceptable to the board and your program.

The CISO should also provide the board with clear, concise, and easy-to-understand reports. Each board is different, and you may expect different items in a report. A Fortune 500 board with global reach may ask a CISO to report on risks across each region of the globe, whereas a board at a small company might just expect the top risks the CISO is addressing and how they are continuously ensuring those risks are mitigated. Tailor reporting to board makeup and needs.

It is obvious that the CISO should be a trusted advisor to the board. The company hired a CISO for that role through interviews, background checks, negotiations, and additional HR processes. There is a certain amount of trust inherent in the hiring process. Given that, the board should be able to rely on the CISO for sound advice on cybersecurity matters.

In addition to these best practices, there are several other things that the CISO and the board can do to improve their working relationship. These include the CISO, who should be willing to educate the board on cybersecurity matters. This may involve using plain language and avoiding technical jargon. It can also involve sending periodicals on governance related to the board, the CISO, and a cybersecurity program.

The board should be willing to invest in cybersecurity. This includes providing the CISO with the resources they need to do their job effectively. Executive management may also be included in these tasks and initiatives.

The board should be supportive of the CISO's efforts. This means backing the CISO's decisions with caution and providing them with the authority they need to do their job.

By following these best practices, the CISO and the board can build a strong working relationship to protect the company from cyber threats.

Here are some additional tips for CISOs and boards to improve their interaction:

- *Be clear about expectations.* The board should set clear expectations for the CISO, and the CISO should be what they need from the board.

- *Be open and transparent.* The CISO should share information with the board in a timely manner, and the board should be willing to listen to any concerns that may arise.

- *Be anticipatory.* The CISO should not wait for a problem to occur before bringing it to the board's attention. The quicker matters are brought to the board's attention, the quicker they can be resolved.

- *Be a team player.* The CISO should be willing to work with the board and any other members of the executive team to develop and implement cybersecurity policies and procedures, as well as initiatives that will help improve the program.

Additionally, the executive team and the board should be willing to work with the CISO and understand his or her concerns related to cybersecurity matters.

By following these tips, the CISO and the board can build a strong working relationship that will help protect the company from cyber threats.

Additional items that are required by a CISO include:

1. *Understanding the Business.* The CISO should have a deep understanding of the business, its mission, and its goals and be able to explain how security fits into the overall strategy and how security can be an enabler for the business versus just a cost center. This will help the board understand the importance of security and make informed decisions.

2. CISOs, as mentioned before, should *Speak in Business Terms.* They should be able to communicate in business terms that the board can understand rather than the technical jargon that they are used to speaking. This means focusing on the risk to the business, the potential impact of security incidents, and the cost of mitigating cyber risk.

3. The CISO should also *Align the Security Objectives With the Corporate Goals* of the organization. This will ensure that security investments and initiatives support the business's strategic goals. Again, this helps the CISO build business cases and be more of a business enabler versus a cost center for the company.

4. CISOs should *Collaborate With Other Departments* such as legal, compliance, risk management, HR, and other areas of the business to have an integrated approach to security. This helps the board see that security is not just an IT issue but a business risk that infects the entire organization.

5. CISOs also *Provide Education and Training* to the organization. CISOs can provide education and training to the board on cybersecurity, including the latest threats, trends, and best practices. This helps the board understand the importance of

security and make informed decisions. Just as the CISO can also provide education and training across the organization, it is important that the board is also educated.

6. The CISO should *Provide Metrics* to the board that demonstrate the effectiveness of the security program, such as the number of security incidents that may occur in a quarter or month, whatever your cadence is, the average time to detect and respond to incidents, the effectiveness of security awareness training, and also the effectiveness of your managed security service provider. This will help the board measure the return on investment in security.

Board Responsibilities

Cybersecurity is a top priority for businesses of all sizes, but it is especially important for Fortune 500 companies and larger companies. These companies have a large amount of sensitive data that is a target for cybercriminals. A data breach at a Fortune 500 or very large company can have a devastating impact, including financial losses, reputation damage, and regulatory fines.

The board of directors of Fortune 500 or large public companies is responsible for ensuring that the company is adequately protected from cyber threats. This includes large, critical infrastructure companies. This means the board must understand the company's cyber risk profile to develop clear cybersecurity policies and procedures through an approval process during committees or board meetings. Monitor the company's cybersecurity posture. Hold management accountable for cybersecurity.

The board should also have regular dialogue, as mentioned before, with the company's chief information security officer about cybersecurity risks and threats. The CISO should be a trusted advisor to the board, and the board should be willing to invest in cybersecurity to protect the company's assets, including shareholder value.

There are many reasons why the board of directors of critical infrastructure companies, large public companies, and any other company should properly govern a cybersecurity program and related cybersecurity risks.

These reasons include the fact that a data breach can lead to significant financial losses for a company. The company may have to pay for the costs of investigating the breach, notifying affected customers, repairing any reputational damage, and, in some cases, paying regulatory fines.

A data breach can also damage the reputation of a company. Customers may lose trust in the company, and the company may find it more difficult to attract new customers as well as retain current customers.

Regulatory fines are also an issue that the board needs to address. In some cases, a data breach can lead to regulatory fines. For example, the European Union's general data protection regulation can impose fines of up to 4% of a company's global annual turnover or twenty million euros, whichever is greater.

A data breach can also disrupt the business. The company may have to shut down its systems and operations for an extended period of time, which can lead to lost revenue, lost productivity, and lost opportunity costs.

In addition to these reasons, the board of directors at Fortune 500 companies and critical infrastructure companies, as well as other-sized companies, should also consider the following factors when governing cybersecurity risk:

- *The Company's Industry.* Some industries are more vulnerable to cyberattacks than others. For example, financial services and healthcare companies are often targeted by cybercriminals due to the amount of privacy records they hold.

- *The Company Size* also has to be considered. Larger companies tend to have more sensitive data, complex IT systems, and intellectual property, making them more of a vulnerable target for cyber threat actors.

- *The Company's Geographic Reach.* Companies that operate in multiple countries may be exposed to different cybersecurity threats depending on the region, and they may not have the same practices that more developed nations have.

The company board of directors should consider all these factors when governing cybersecurity risk. By doing so, the board can help protect the company from cyber threats and minimize the damage a breach could cause.

In addition to the above, here are some specific actions that the board of directors at large global critical infrastructure and public companies and smaller companies can take to properly govern cybersecurity risk.

1. The board Needs to *Set Clear Expectations for Management*. The board should set clear expectations regarding cybersecurity, which include establishing clear goals and objectives, developing

and implementing appropriate policies and procedures, and allocating adequate resources.

2. The board should also *Hold Management Accountable for Cybersecurity,* including the CISO. This means reviewing the company's cybersecurity posture regularly, staying in touch with the CISO, understanding the cyber risk that the CISO and security team are mitigating, and taking any necessary action to help mitigate the risk.

3. The board should *Get Regular Updates From the CISO* about these risks. This will help the board stay informed and make informed decisions about cybersecurity and the practices that the company is involved in.

4. The board, as well as the company, should *Invest in Cybersecurity* to protect company assets. This includes investing in people, technology, training, and processes that go along with cybersecurity practices.

5. Boards should be *Active in Addressing Cybersecurity Risks,* just as the CISO should learn what risks might rise to the level of materiality and should bring those risks to the board's attention as well as executive management's attention. This means anticipating threats and taking steps to mitigate them.

As a recommendation, I suggest any board, whether public or private, global or smaller, take a look at the NACD 2023 Director's Handbook on Cyber Risk Oversight to learn about more ways to govern a cyber program and minimize residual risks. There are some highly effective principles in the Handbook that can help a company reach its cyber resilience goals.

By taking these actions, the board of directors of Fortune 500 companies, large public companies, and companies of all sizes, including critical infrastructure companies, can help to protect the company they assist in governing from cyber threats and minimize the damage that a data breach could cause.

CHAPTER 7

The Future Is Here

It is June 2024; both of the major political parties in the United States have decided on their representatives for the upcoming November 2024 presidential election. The continuous news cycle talks of social media meddling by foes of the United States that are trying to sway the outcome. CISA, the FBI, and the CIA are tracking the threat actors who are propagating false narratives to stir up American tension.

Meanwhile, a splinter group of a nation-state (Iran) threat actor has broken off, wanting to cause catastrophic damage to America's critical infrastructure, utilizing a multi-state attack based on their ideology of hate towards the Western world. The plan is a large-scale cyberattack that is to cause serious disruption to Wall Street operations and cripple trading for days or weeks. Further, the plan includes infecting a major Houston area refinery operation that is to cause a major explosion and lead to a deadly pollutant being sent throughout the Houston region via Gulf Coast winds. A third attack is a plan to disrupt Washington, D.C., public transportation (Washington Metropolitan Area Transit Authority) by causing a malfunction in the rail system, diverting trains and subways to collide with one another in a near-simultaneous event.

Luckily, a threat intelligence analyst spotted this plot, which was initially thought to be part of a nation-state (Russia) group acting on behalf of Russia. The Advanced Persistent Threat actor group was discovered by the analyst to be a splinter group of another Iranian APT and disagreed on the scale of cyberattacks they were currently perpetrating across the globe; they decided to split from them and started plotting to cause devastating damage to critical infrastructure with the intent of causing the loss of human life. She had been monitoring the groups for the past year and detected the tension between each and differing ideologies to the extent the cyber war could be pushed.

While the above scenario is completely fictitious, it is plausible to all experts in cybersecurity and is a potential scenario many of us think about. Cyber warfare and the lack of cybersecurity excellence are serious threats to individuals, businesses, and governments around the world. Cyberattacks are becoming more sophisticated and frequent and can have a devastating impact. In order to defend against the current state of cybersecurity, international laws and United States laws need to be strengthened. Here are some specific recommendations.

International laws should be comprehensive, and an international treaty on cybersecurity should exist. This treaty would provide a framework for cooperation between countries on cybersecurity issues, such as sharing information about threats, developing standards, and enforcing laws. International laws should also consider additional laws, such as the United Nations Charter and the Convention on the Law of the Sea, that need to be updated to address the challenges of cyber warfare. United States laws should encompass a comprehensive cybersecurity law that would provide a framework for cybersecurity in the United States, including setting standards for businesses and government agencies and providing funding for research and development.

We have already seen some of these initiatives take place in the United States. Additional measures related to laws might include strengthening the existing United States laws on cybercrime. These laws, such as the Computer Fraud and Abuse Act, need to be updated to address the challenges of cybercrime.

In addition to these specific recommendations, there are a number of other things that can be done to improve cybersecurity. Increase awareness of cybersecurity threats. Individuals and companies should be aware of the risk of cyberattacks and how to protect themselves. There are numerous organizations that can help with this, such as InfraGard or information sharing and analysis centers (ISACs).

Improve the education related to cybersecurity within businesses, the government, and schools. Businesses and government agencies should provide their employees with cybersecurity education and training. Schools K through 12 and universities should be providing adequate training modules that can help students understand the risks related to cybersecurity.

We can also invest more in cybersecurity research and development at a holistic level. As a country and businesses, we must continue investing in research and development to develop innovative technologies to defend against cyberattacks. This includes technologies we have mentioned before, such as blockchain, quantum computing, and many others. Cyber warfare is a growing threat to the world, and it is important to consider the ethical implications of this new form of warfare.

There are several ethical considerations that should be taken into account when it comes to cyber warfare. These include proportionality. Cyberattacks should be proportionate to the threat to which they are responding. For example, if a country is attacked by a cyberattack that causes minor damage, it would not be ethical to attack it with a cyberattack that causes major damage across the country.

This is much easier said than done. Many countries will ratchet up attacks. Cyberattacks should not target civilians or civilian infrastructure. For example, many do not believe it ethical for a nation-state to indiscriminately attack a power grid that provides electricity to civilians in the middle of winter or summer, when climates can be harsh. Yet we see this happening day in and day out. These ethics need to be at the forefront of a cybersecurity treaty. While I believe warfare and even ratcheting up that warfare against the enemy's critical military targets is important, most wars are related to government disputes or differing ideologies with little civilian involvement.

Responsibility for how nation-states should act with state cyber actors and cyber warriors. If a state's threat actors and cyber warriors launch a cyberattack that causes accidental or intentional damage, the state should be held accountable. There needs to be some level of transparency across cyber warfare capabilities and activities; it is often done in physical warfare as a means of deterrence. This would also help to build trust and cooperation between countries and deter cyberattacks against countries and allies.

These are general ethical considerations. There are also a number of specific ethical considerations that need to be taken into account when it comes to attacks on our critical infrastructure as well as on each country. Cybersecurity attacks that cause widespread harm, such as to power grids, water treatment plants, or other critical infrastructure, need to be considered. Ethical boundaries are particularly important to establish when it comes to cyber warfare. Establishing these boundaries can help ensure that cyber warfare is used responsibly and ethically.

As the 21st century continues to advance and technology becomes more sophisticated and entrenched in our lives, it has created a dramatic shift in the digital landscape. "Cyber" has become the fifth domain of warfare, joining land, sea, air, and space; cyberspace is the fifth domain. But unlike traditional battlefields, the digital terrain is nebulous, relentlessly evolving, and immensely challenging to navigate. The ongoing cyber warfare that nations engage in profoundly influences how our future will unfold.

A digital cold war is stirring. If nations refrain from full-blown physical confrontations in the decades to come, we might see a digital cold war unfold. Just as the U.S. and U.S.S.R. engaged in decades of proxy wars, space races, and nuclear standoffs, nations of the future might compete for cyber supremacy. We will witness silent cyberattacks, espionage, counterespionage, and nations bolstering their digital arsenals. The long-term effects will be a proliferation of cyberweapons, similar to nuclear proliferation but much easier to acquire and operate. There will be increased emphasis on cyber-deterrence strategies, much like with physical weaponry. Covert operations will become the norm and the preferred mode of confrontation, with no entity taking responsibility.

Infrastructure will become more fragile and susceptible to attacks because it is a soft target, rich with vulnerabilities, and has a higher loss when it comes to damage. As nations increase their reliance on digital systems, from electricity grids to water supply networks, these become vulnerable targets. Cyberattacks might aim at disrupting daily life, causing not just economic harm but also potentially leading to physical catastrophes. Companies might revert to maintaining some analog or isolated systems as backups. We may see an enhanced focus on infrastructure security leading to increased costs, either through advanced implementations or penalties or fines due to a lack of security, with those costs being passed on to individuals in the form of higher taxes or higher costs for those services. Then you have the ever-evolving game of breach and patch, as attackers find new vulnerabilities and defenders seek to fix them.

Another effect is a shift in power dynamics. Smaller nations or even non-state actors could potentially challenge superpowers in the cyber realm. Digital warfare can level the playing field, allowing those with technical knowledge to wield disproportionate power. Traditional notions of power and influence will be challenged. Non-state actors like hacktivist groups or multinational corporations may gain significant geopolitical influence. Diplomatic dialogues may often involve negotiations over cyber armistices.

There are economic repercussions to consider. The economy of the future will be deeply intertwined with the digital realm. Sustained cyber warfare could lead to a loss of investor confidence, disruptions in supply chains, and economic crashes. We are already seeing some of the ramifications of the many breaches by Fortune 500 companies and critical infrastructure organizations that have caused serious disruption in our supply chains. To counter this, nations might invest heavily in creating isolated, self-reliant digital economic zones. Insurance markets will continue to increase prices to be insured and further restrict those who can be insured due to a lack of a strong cyber posture. Additionally, new professions will be created that are centered around cyber warfare mitigation and economic recovery. These are unfolding now with the creation of new government agencies and enhanced regulations.

Social impacts will permeate through all of this. The psychological impacts of living under constant cyber threats, whether it be loss of identity, loss of finances, or worse, cannot be underestimated. People might begin to distrust digital platforms, leading to a more fragmented society. One might consider a potential return to offline modes of communication and transactions. There will be an increase in mental health issues related to digital trust and privacy concerns, much like we saw during the height of the COVID-19 epidemic from 2020 to 2021, not to mention educational impacts. Shifts in culture and what we value, minimizing digital usage, could sprout among certain individuals.

Cyber warfare has expanded the theater of conflict beyond traditional battlefields. The combination of evolving technologies like AI and drones, coupled with socioeconomic disparities, produces a variety of futuristic attack scenarios.

When it comes to critical infrastructure, a nation-state, with the help of AI, can easily identify vulnerabilities in another country's electrical grid, transportation systems, supply chains, or other areas that expand throughout a country. Threat actor groups, with the facilitated scale and coordination of an attack through the use of AI, have the capability and motive to commence an attack, causing widespread blackouts and affecting hospitals, transportation, and other essential services.

Threat actors may use a sophisticated AI-driven algorithm to breach the security of major financial institutions or stock exchanges, causing panic, market crashes, or draining citizens' bank accounts. This type of attack is most plausible when you consider the socioeconomic disadvantages of less developed and smaller countries, much like the physical warfare today.

Another scenario that is rapidly developing is the use of drones. Drones equipped with hacking tools are flown into enemy territories. These drones then locally breach secure networks, spying on, disrupting, or damaging critical infrastructure without the need for remote access.

Drones, like other technologies, can bypass traditional network defenses by attacking locally. Drones, as we are witnessing today in the Ukrainian defenses against Russia, can be used for physical sabotage as well, combining digital and physical warfare.

As mentioned before, threat actors are compromising software or hardware during the manufacturing process or other processes related to operations, much like we have seen in recent breaches such as SolarWinds Orion. When these compromised products are integrated into larger systems or customer operations, they become Trojan horses, providing backdoor access to critical systems. The use of AI can help attackers identify the most impactful point in the supply chain to compromise. It helps automate the process of seeking out vulnerabilities once a system is compromised.

While we are seeing nations deploy remote-controlled drones frequently now, in the future, a nation's military may deploy autonomous drones or robot soldiers as part of warfare. An enemy state then hacks into these systems, turning them against their owner or causing them to malfunction. This represents a dual-edged sword: while they can enhance military capabilities, they can also become liabilities if compromised. Remarkably similar to today's cybersecurity tools, applications, and other things at our disposal.

Other scenarios that may play out are economic disadvantage exploits where wealthier nations with advanced cyber capabilities launch attacks on poorer nations, exploiting their lack of resources to defend against sophisticated attacks. These could be for gaining political leverage, extracting resources, or espionage. Data manipulation and espionage are utilized when AI-driven bots infiltrate databases, altering information subtly. Over time, these alterations cause chaos in financial records, medical databases, or electoral rolls.

The combination of AI, drones, and socioeconomic factors in cyber warfare present a multifaceted challenge. While technology offers enhanced capabilities, it also introduces vulnerabilities. It is crucial for nations, regardless of their economic status, to understand these scenarios and work collaboratively to establish norms, defense strategies, and education campaigns to mitigate risks.

The contours of the future are shaped by the actions we take today. Cyber warfare reflects our digital dependence and brings new challenges and opportunities. As the world grapples with its implications, it's vital for nations, organizations, and individuals to understand the long-term scenarios to better prepare for and navigate the intricate web of the digital future.

From the Desk of a CISO: A CISO's Perspective

Cybersecurity Collaboration Featuring Michael Farnum

Michael Farnum Bio: Michael Farnum is an advisory CISO and has been in the IT and information security fields since 1994. During this time, Michael has held numerous senior cybersecurity consulting, architecture, and leadership roles in various industries, such as healthcare, staffing, and software development. Michael has spent the last several years working for cybersecurity solutions providers in the consulting, application security, and reseller spaces, where he has helped numerous organizations assess and build their security and compliance programs.

In 2010, Michael founded HOU.SEC.CON, a non-profit cybersecurity conference in Houston with more than 1,000 attendees annually. In addition to organizing a security conference, Michael often speaks at conferences such as RSA, Infosecurity Europe, and several BSides events, and he is also a prolific cybersecurity podcaster.

Before his career in IT and security, Michael served in the U.S. Army, drove an M1A1 main battle tank in Desert Shield/Storm, and spent 2+ years as a land surveyor.

"Our method was to develop integrated products, and that meant our process had to be integrated and collaborative" - Steve Jobs, co-founder and former CEO of Apple

"When I was a kid, there was no collaboration; it's you with a camera bossing your friends around. But as an adult, filmmaking is all about appreciating the talents of the people you surround yourself with and knowing you could never have made any of these films by yourself" - Steven Spielberg, 3-time Academy Award-winning Director

Collaborating With Trusted Partners is Key to a Resilient Cyber Program

Critical infrastructure represents the backbone of any nation's economy, health, safety, and security. We rely on power grids, water supply stations, refineries, transportation systems, healthcare facilities, and financial institutions; these vital assets are pivotal for the day-to-day operations of a society. As has been stated before, these systems' technological complexity and connectivity continue to grow. Along with that is the challenge of ensuring the security of these assets. In such a critical, high-stakes environment, collaboration with a trusted technology partner emerges as a strategic advantage and an imperative.

"It's important to build trust with partners not only before but also during collaborations. This helps foster long-term connections." - Michael Farnum

There are many ways to do this: Engage partners across innovation processes. No single organization can possess comprehensive, expert knowledge of every technological domain or security challenge. Trusted technology partners, beyond ones that seem to only want to sell you a product, often bring niche expertise or a wide scope of best practices, specialized tools, and years of experience to their clients. Their insights can accelerate the process of identifying vulnerabilities, developing mitigations, and enhancing the overall security posture of critical infrastructure.

"Thinking from the context of a true partner, that is somebody who is going to offer value outside of just a product they want to sell you. There has to be a lot of pre-sales value" - Michael Farnum

CISOs should look for trusted partners who are as curious as they are and interested in how the company's security program has been developed, what strategies are in place, what architectures have been designed, and genuinely want to understand your business.

The digital threat landscape is not static; it evolves rapidly, with adversaries continually devising new techniques and strategies. A trusted technology partner dedicated to staying abreast of these changes can ensure that protective measures are not just reactive but also proactive, anticipating future threats and reinforcing systems accordingly. Protection of critical infrastructure is resource-intensive, requiring both technological and human capital.

When a CISO is able to collaborate with a trusted technology partner, that partner can bring in additional resources, whether it's advanced analytics tools, threat intelligence information, or specialized cybersecurity personnel, allowing for a more holistic and effective cybersecurity strategy without further compounding the constraints on internal resources.

Another area where trusted partners can provide an exponential amount of value is through incident response. When security incidents occur, swift and coordinated response measures can mean the difference between minor disruptions and major catastrophes.

"It's a question that I am often getting from CISOs now. It used to be only the smaller companies interested in managed services. Now, I am seeing a lot of enterprises asking about managed services. It may not be the whole program, but they are asking about specific aspects of their program." - Michael Farnum

Trusted technology partners can provide crucial support during cybersecurity incidents, offering both technical solutions derived from supporting many clients in times of crisis and strategic guidance in all phases of incident response, ensuring that incidents are mitigated effectively, and the disruption is minimized.

Trusted technology partners often have a broader view of the industry, interacting with diverse clients and facing many challenges.

"When they work with technology partners, I believe their success hinges on efficient collaboration with the technology suppliers and consultants who help them streamline their operations, launch exciting new products and services, and stay relevant in an increasingly tech-driven world." - Michael Farnum

This exposure can lead to innovative solutions, novel methodologies, and best practices that can be instrumental in fortifying critical infrastructure.

"Being able to leverage enough anecdotal evidence to make that factual, then combining that with what is innovative and coming down the pipe is key. Then the partner has the relationships to look at start-up companies, and the roadmaps of current companies help a client with their program" - Michael Farnum

With growing regulatory scrutiny on the security of critical infrastructure, organizations face the dual challenge of ensuring robust security while complying with complex regulations.

"Another value that a partner can bring to the relationship is what current risks are top of mind throughout their client ecosystem. Two that I am hearing from now are the continuous new national and state privacy and cybersecurity regulations and third-party risk management" - Michael Farnum

Trusted technology partners can assist in navigating this regulatory maze, ensuring that systems are compliant while minimizing risks.

"Value can not only come from the engineering side but can also come from the pre-sales team in the form of budgeting, putting a business case together, pricing, and a number of other things" - Michael Farnum

Collaborating with a respected technology partner can bolster confidence among stakeholders, from customers and investors to regulators and the general public. It sends a clear message that the organization is committed to the highest security standards and is taking proactive measures to safeguard critical assets.

"Being able to know what is included in your technical environment. What maturity level does the client want to achieve? Helping client CISOs with board presentations so the CISO is not just out there on his or her own having to build the presentation in a vacuum." - Michael Farnum

As the complexities of critical infrastructure and its associated threats multiply, the role of a trusted technology partner becomes increasingly pivotal. Such partnerships amplify capabilities, enhance resilience, and offer a more adaptive and agile approach to security challenges. In the high-stakes realm of critical infrastructure protection, where the repercussions of failure can have wide-ranging societal impacts, the value of strong collaboration cannot be overstated.

CHAPTER 8

America's and our Allies' Path Forward

As we have seen in the previous chapters, the current state of cybersecurity is a serious threat to individuals, businesses, and governments worldwide. Cyberattacks are becoming more sophisticated and frequent and can have a devastating impact. The technological leaps we have made in the past decade have transformed how wars are fought. Cyberspace has become the fifth domain of warfare, joining land, sea, air, and space. And just as in these other domains, the cyber landscape is unpredictable and volatile.

The United States and our global allies are facing a growing cybersecurity crisis. Critical infrastructure such as power grids, water systems, and transportation networks is increasingly vulnerable to cyberattacks. These attacks can have devastating consequences, causing widespread power outages, water shortages, transportation disruptions, and even the health of employees at many of these critical infrastructure companies. The threat of cyberattacks is not limited to infrastructure. Businesses of all sizes are also at risk. In 2022, the average cost of a data breach was estimated to be $4.24 million. These breaches can lead to the loss of sensitive customer data, financial losses, damage to a company's reputation, and intellectual property losses.

The United States government is in a cybersecurity crisis. In 2018, the President signed an executive order that created the Cybersecurity and Infrastructure Security Agency, otherwise known as CISA. CISA is responsible for coordinating the federal government's efforts to protect critical entities. However, more needs to be done. The United States government needs to invest in research and development. Much like we created the Peace Corps of the past, the United States government, and our allies need to work with industries and educational institutions to further create the Cyber Corps to help defend our nation against future threats and work with businesses to help them improve.

Cyber threat vectors continue to grow, and the evolving landscape of cyberattacks and cyber warfare becomes increasingly complicated. In order to defend against the current and future state of cybersecurity, the United States and our allies should develop cyber peace treaties, establish unified cyber commands and bodies, and invest in cyber education throughout all levels of school, from K–12 to postgraduate. Laws need to be strengthened across the United States and among our allies.

Our future state looks daunting. Our critical infrastructure continues to be a soft target. Cyber warfare has expanded the theater of conflict beyond traditional battlefields. The combination of evolving technologies like AI and drones, the ease and availability of attack techniques, and the proliferation of technology being dispersed to and into everything, coupled with macroeconomic factors and more disparities, defines a world where cyber risks are exponentially increased.

As much as we have challenges, we also have opportunities. America's approach must be dynamic, forward-thinking, and collaborative. The future of national security will depend on traditional military strength and the ability to navigate, defend, and influence in the digital realm. By recognizing the significance of cyber threats and proactively addressing them, America and our allies can safeguard our critical infrastructure and global interests.

Opportunities relate to strengthening international and U.S. laws. The United States should lead in developing a comprehensive international treaty on cybersecurity. A treaty that provides a framework for cooperation between countries on cybersecurity issues, such as sharing information about threats, developing standards, and enforcing laws, should be invoked. Strengthening existing international laws that apply to cyber warfare or include cyber warfare should be done.

In the United States, we need to pass a comprehensive cybersecurity law. We are starting to see movement in this area, with each administration creating executive orders and new government bodies. A U.S. law would include standards for businesses and government agencies, provide funding for research and development at all levels, and increase penalties and the comprehensiveness of current laws on cybercrime. Incentives, much like we do in other areas of business, could be provided for those who meet and/or exceed these standards.

Ethical considerations should be taken into account when defending against cyberattacks as well as engaging in cyber warfare. Questions about proportionality, collateral damage, and the responsibility of state actors need to be answered and agreed upon. The United States and our allies (NATO and other countries) should determine what is out of bounds, the collateral damage that could occur to country citizens, proxy groups vs. actual governments vs. military governments.

Cyberattacks should be proportionate to the threat they are responding to, and deterrence should be at the pinnacle of engagement strategies. Much like we and our allies lead the world in ethical behaviors in physical wars, we should not target civilians or civilian infrastructure in cyber warfare where boundaries exist. To build trust, responsibility and transparency should be included, such as holding those accountable who step outside the boundaries of treaties and further cooperation with allies to the extent prudent. Additional guardrails related to data privacy, due process, and how warfare affects basic human rights should be considered.

The United States and our allies need to increase our investment in cybersecurity at all levels—teaching institutions, businesses, and government. Research and development, which we are seeing a lot more of, will be key to being a leader in cyber defense and using those defenses as a warning. Investments should be used to develop new technologies—autonomous, self-learning, swarming, and others—to defend against cyberattacks and improve understanding cyber threats.

More cooperation between government and industry, such as bodies like InfraGard, is needed. The United States and our allies should improve cooperation between government and industry to address the threat of cyber warfare, including additional means of sharing information about cyber threats, developing joint cybersecurity standards and exercises, and enforcing international cybersecurity laws. Cooperation should include a significant education and awareness component covering risks, protection baselines, and reporting. We have these mechanisms in place now, but the average citizen does not know them. The teaching should begin in K-12 classrooms.

We have seen the beginning of some of these initiatives, including critical infrastructure guidance, the creation of a government national cybersecurity strategy, and the establishment of DHS's CISA, the Department of Homeland Security Cybersecurity and Infrastructure Security Agency, which have been great first steps. More guidance and practical recommendations to critical infrastructure owners and operators on how to improve their cybersecurity posture can be enforced, including mandating that manufacturers provide a baseline level of training and require operators to be certified in best practices for securing systems, operational technology systems, and physical security operations of systems.

Our national cybersecurity strategy should now be taken and disseminated across all sectors of critical infrastructure and business. The next phase of a comprehensive strategy should outline the private sector's involvement. Not only for large businesses but for small businesses as well, a tailored approach according to size, complexity, and data custodian responsibilities is needed.

Though we have established CISA, we should establish something similar on a volunteer basis and across states. Similar to many of our national, state, and regional associations, a cybersecurity volunteer center of excellence is required. This body could provide knowledge on best practices in protection, educational opportunities, response guidance, sharing of information, and policy/standard guidance. Think of it as a volunteer cyber peace corps.

Cyber warfare, continuous attacks on our critical infrastructure, and the cyber arms race are not knocking on our door; they are in our living rooms. The decentralized nature of the internet and cyberspace means America cannot rely solely on its traditional strengths (like geographical distance) for defense. However, the U.S. and our allies have the potential to harness our vast technological and financial resources, innovative private sectors, and rigorous academic research institutions to take the lead in cybersecurity and stand ready to disrupt attacks on our critical infrastructure and data.

We can no longer collectively stand by with our hands in our pockets, stating that it is someone else's problem, that it will never happen, or waiting for another country to solve this crisis. Our enemy may solve it before us, and that complacency would have dire consequences. Heavy is the head that wears the crown. The United States, as a whole, has always, since our inception, worn it with diligence and great responsibility. Cyber warfare and leadership in cyberspace should be no different.

CONCLUSION

Is a Cyber War Really Here?

Recently, on August 31, 2023, an article was published that related to NATO (North Atlantic Treaty Organization) pondering the use of Article 5 in cases of cyberattacks. Article 5 essentially provides that if a NATO Ally, one of the thirty-one countries, is the victim of an armed attack, each and every other member of the Alliance will consider this act of violence as an armed attack against all members and will take the actions it deems necessary to assist the Ally attacked.

Given this, the number of cyberattacks against the United States and allies' critical infrastructure, and many other ongoing activities related to cyber incidents and warfare, my opinion would be that, yes, a cyber war is really here. Several factors are driving the cyber war, most importantly the increasing importance of information technology in modern societies, the proliferation of technology throughout our human ecosystem and the growing interconnectedness of it, and the overall difficulty of its provenance.

A cyber arms race has existed since the early 1990s, when nation-states began to develop and deploy cyber weapons. These early weapons were relatively simple, but they were still capable of causing severe damage. In the years since, the sophistication of cyber weapons has increased dramatically, and they are now capable of causing widespread disruption and destruction. Recently, we have seen both Ukraine and Russia use cyber as a weapon in the ongoing Russian and Ukrainian war.

While there has not been a country to come out and officially declare a cyber war, American and global infrastructure and other high-value targets are constantly under cyberattack. We see several nation-state-backed groups, whether government-based, government-backed, or national threat actor groups, perpetrating these attacks against U.S. and global critical infrastructure, including U.S. government agencies and government officials.

Nation-states, criminal organizations, and other threat actor groups have launched numerous cyberattacks against critical infrastructure in the past, including attacks against power grids, water systems, and other essential services. Many of these attacks have caused significant disruption and damage, including the theft of personally identifiable data, and resulted in serious interference in providing health care, but attacks have not risen to the level of a full-blown cyber war. With NATO pondering Article 5 in the case of cyberattacks, that may be changing.

A silent cyber war has been going on for a while, whether through the use of proxies, government-backed groups, or threat actors after economic gain. Other signs that cyber war may be escalating are the use of cyber capabilities as weapons, the continuation of state-sponsored cyber activities, more sophisticated tactics and techniques, coordination among specific state groups, and accelerating attacks on critical infrastructure to include harm to humans.

In a cyber war, just like in a physical war, warfare would likely be widespread and systematic, with multiple targets being attacked simultaneously, potentially across multiple countries or regions. Physical targeting could be done against adversaries where attacks are being launched from high-value targets to get the message of deterrence across, and strong geopolitical tensions may crop up as a means of exerting political or economic pressure.

Important Lessons to Remember

- *Cyber warfare can level the playing field more than physical warfare.*

   We have seen in the war against Ukraine by Russia that Russian cyber capabilities have been less successful than most predicted. Jon Bateman, a senior fellow in the Technology and International Affairs Program at the Carnegie Endowment for International Peace, put it best when he said that "exceptional defensive efforts by Ukraine and its partners" have resulted in Russia's low success rate. After years of being under attack by Russia, Ukraine seems to have fared well. As they say, practice makes perfect. While war is anything but perfect, Ukraine has withstood Russia's best, which, at the time of the conflict's beginning, was thought to be among the most capable in the world.

- *Cyber Risks are Everyone's Risks; We are All Targets.*

   Individually, whether you work at a small business, a global corporation, or a government entity, cyberattacks are constantly targeting us. There is no hiding from it. Never let another crisis go to waste. We can learn from the many attacks that have occurred recently and what were some of the factors that allowed the breach. Basic stuff: unpatched systems, not using MFA, not testing your cyber resilience, and on and on. We should all be engaged in mitigating cyber risks and play a part in their reduction. At a minimum, corporate and government employees should be aware of the risks.

- *The Board and Tone at the Top Matters.*

Boards and company executives should be involved in the decision-making process and have active oversight of cybersecurity. An article in the Harvard Business Review stated, "The board tends to strategize about ways to manage business risks; cybersecurity professionals concentrate their efforts at the technical, organizational, and operational levels." Often, CISOs and the board, as well as business executives, do not speak the same languages. The common language is risk and how to convey the level of risk. The board, company executives, and the executive leadership team should ensure the company is as prepared as demonstrably feasible to withstand a cyberattack based on the company's complexity, risk tolerance, and materiality levels.

- *Cyber Could Not Be Any Cooler*

We got the swag; we got the swag! Cyber is a fun field, and SASE (sassy)—the cyber nerds will get that. Conferences, happy hours, events, speaking opportunities, fellowship, writing—it is all there. People from all walks of life are entering the cyber field. And it is not just us cyber nerds hammering away on keyboards. Just like any other industry, we have marketing, engineering, strategy, learning and development, sales, tech gods, innovators, policy development and planning, analysts, finance, and many other disciplines in the cyber industry. AI (Artificial Intelligence), Blockchain, Cloud security, IoT (Internet of Things) security, ICS (Industrial Control System) security, network security, endpoint security, security architecture, yada, yada, yada. Need I say anything about the longevity of cyber? Jump in, hold on, and be prepared for change.

- *Cyber Resilience is the New Defense in Depth*

    Cyber resilience, an organization's ability to anticipate, withstand, recover from, and adapt to adverse conditions, stresses, attacks, or compromises on systems that use or are enabled by cyber resources, has replaced defense in depth. Whereas defense in depth primarily dealt with technology and the layers of technology needed to be alerted to or thwart a cyberattack, cyber resilience as a built-in process also emphasizes processes and people. Cyber resilience is also about adapting to adversaries and the continuously evolving threat landscape and techniques used.

In times of crisis, I often wonder, *What would he do?* or *What would she do?* And even *What would Jesus do?* Not in my own crisis, but monumental ones on a global scale, like COVID-19, the financial crisis in 2008, 9/11, or any other globally significant event we have encountered in our lifetime. We are facing what I believe history will remember as a technological or cyber crisis. Cyber warfare is happening all around us. Whether it is a war against you and/or your child, your company, or your country, now is the time to get involved.

Ask yourself, *What would Lincoln do now?* (substitute your favorite president, authoritative figure, or leader.)

May this read have been worth your journey. Godspeed and all my best.

Bibliography

Confessions and Acknowledgements:

Reagan Approved Plan to Sabotage Soviets, David E. Hoffman, https://www.washingtonpost.com/archive/politics/2004/02/27/reagan-approved-plan-to-sabotage-soviets/a9184eff-47fd-402e-beb2-63970851e130/, February 27, 2004

160 Cybersecurity Statistics 2023, Nivedita James, https://www.getastra.com/blog/security-audit/cyber-security-statistics/, August 4, 2023

Introduction:

2023 Cybersecurity Almanac: 100 Facts, Figures, Predictions, and Statistics, Cybercrime Magazine, Steve Morgan, https://cybersecurityventures.com/cybersecurity-almanac-2023/, May 24, 2023

Cybersecurity Trends & Statistics for 2023: What You Need to Know, Chuck Brooks, https://www.forbes.com/sites/chuckbrooks/2023/03/05/cybersecurity-trends--statistics-for-2023-more-treachery-and-risk-ahead-as-attack-surface-and-hacker-capabilities-grow/?sh=5d314b1119db, March 5, 2023

The Clorox Company, Press Releases, Clorox Provides Preliminary Q1 Financial Information and Operations Update, https://s21.q4cdn.com/507168367/files/doc_news/Clorox-Provides-

Preliminary-Q1-Financial-Information-and-Operations-Update-2023.pdf, October 4, 2023

Updated: Clorox Cyberattack to Cost Up to $593 million, Dennis Scimeca, https://www.industryweek.com/technology-and-iiot/article/21274431/the-clorox-co-recovers-from-severe-cyberattack, October 5, 2023

Clorox, reeling from cyberattack, expects quarterly loss, Savyata Mishra and Zeba Siddiqui, https://www.reuters.com/business/retail-consumer/clorox-expects-quarterly-loss-hit-cyberattack-2023-10-04/, October 4, 2023

Caesars Entertainment says social-engineering attack behind August breach, David Jones, https://www.cybersecuritydive.com/news/caesars-social-engineering-breach/695995/, October 9, 2023

Caesars sheds more light on ransomware-related data breach, SC Staff, https://www.scmagazine.com/brief/caesars-sheds-more-light-on-ransomware-related-data-breach, October 12, 2023

Caesars Data Breach Saw Hackers Steal Over 41,000 People's Data, James Laird, https://tech.co/news/caesars-data-breach-state, October 12, 2023

The chaotic and cinematic MGM casino hack, explained, Sara Morrison, https://www.vox.com/technology/2023/9/15/23875113/mgm-hack-casino-vishing-cybersecurity-ransomware, October 6, 2023

Casino giant MGM expects $100 million hit from hack that led to data breach, Zeba Siddiqui, https://www.reuters.com/business/mgm-expects-cybersecurity-issue-negatively-impact-third-quarter-earnings-2023-10-05/. October 5, 2023

The MGM Resorts Attack: Initial Analysis, Andy Thompson, https://www.cyberark.com/resources/blog/the-mgm-resorts-attack-initial-analysis, September 22, 2023

MGM Facing Class Action Over 10-Day Cyberattack in September 2023, Kelsey McCroskey, https://www.classaction.org/news/mgm-facing-class-action-over-10-day-cyberattack-in-september-2023, September 27, 2023

MGM Resorts confirms hackers stole customers' personal data during cyberattack, Carly Page, https://techcrunch.com/2023/10/06/mgm-resorts-admits-hackers-stole-customers-personal-data-cyberattack/, October 6, 2023

Inside the Ransomware Attack That Shut Down MGM Resorts, Suzanne Rowan Kelleher, https://www.forbes.com/sites/suzannerowankelleher/2023/09/13/ransomware-attack-mgm-resorts/?sh=bae92695f384, September 13, 2023

MGM Resorts Cyberattack Stymies Slot Machines, Check-Ins, Katrina Manson and Jamie Tarabay, https://www.bloomberg.com/news/articles/2023-09-11/mgm-resorts-says-it-shut-down-some-systems-following-cyberattack?leadSource=uverify%20wall, September 11, 2023

'Scattered Spider' Behind MGM Cyberattack, Targets Casinos, Becky Bracken, https://www.darkreading.com/attacks-breaches/-scattered-spider-mgm-cyberattack-casinos, September 14, 2023

Chapter 1:

Infosecurity Magazine, James Coker, Deputy Editor, Infosecurity Magazine https://www.infosecurity-magazine.com/news/cybersecurity-workforce-gap-grows/, October 20, 2022

CSO Online, Jeff Robbins, Practice Director, Security/Wireless, for Business Communications, Inc. (BCI), Cultivating a New Generation of Cyber Professionals, https://www.csoonline.com/article/571681/cultivating-a-new-generation-of-cyber-professionals.html, November 21, 2021

National Cyber Workforce and Education Strategy, The Office of the National Cyber Director (ONCD), Acting Director of ONCD, Kemba Walden, Preparing Our Country for a Cyber Future, https://www.whitehouse.gov/oncd/preparing-our-country-for-a-cyber-future/, July 31, 2023

White House, President Joseph Biden, 2023 National Cybersecurity Strategy, https://www.whitehouse.gov/wp-content/uploads/2023/03/National-Cybersecurity-Strategy-2023.pdf, March 2023

Chapter 2:

Air Force F-22 Fighter Program, EveryCRSReport.com, https://www.everycrsreport.com/reports/RL31673.html, Jan 6, 2005 – July 11, 2013

World Economic Forum, The Global Risks Report 2023, https://www3.weforum.org/docs/WEF_Global_Risks_Report_2023.pdf, January 2023

10.5 trillion reasons why wee need a united response to cyber risk,Forbes, Carmen Ene, https://www.forbes.com/sites/forbestechcouncil/2023/02/22/105-trillion-reasons-why-we-need-a-united-response-to-cyber-risk/?sh=3a28364e3b0c, Feb. 22, 2023

SEC Proposes Rules on Cybersecurity Risk Management, Strategy, Governance, and Incident Disclosure by Public Companies. U.S. Securities and Exchange Commission, https://www.sec.gov/news/press-release/2022-39, March 9, 2022

SEC Adopts Rules on Cybersecurity Risk Management, Strategy, Governance, and Incident Disclosure by Public Companies, U.S. Securities and Exchange Commission, https://www.sec.gov/news/press-release/2023-139, July 26, 2023

Progress Software Releases Security Advisory for MOVEit Transfer Vulnerability, Cybersecurity & Infrastructure Security Agency

https://www.cisa.gov/news-events/alerts/2023/06/15/progress-software-releases-security-advisory-moveit-transfer-vulnerability, June 15, 2023

MOVEit mass exploit timeline: How the file-transfer service attacks entangled victims, Cybersecurity Dive, Matt Kapko, https://www.cybersecuritydive.com/news/moveit-breach-timeline/687417/, July 14, 2023

Resolution adopted by the General Assembly on 23 December 2015, United Nations, General Assembly, https://documents-dds-ny.un.org/doc/UNDOC/GEN/N15/457/57/PDF/N1545757.pdf?OpenElement, December 30, 2015

Resolution adopted by the General Assembly on 22 December 2018, United Nations, General Assembly, https://documents-dds-ny.un.org/doc/UNDOC/GEN/N18/465/01/PDF/N1846501.pdf?OpenElement, December 22, 2018

Chapter 3:

2022 Unit 42 Network Threat Trends Research Report: Insights into newly observed attacks in the wild, Palo Alto Networks, https://start.paloaltonetworks.com/unit-42-network-threat-trends-report-2022, (n.d.)

Everything You Need To Know About BlackCat (AlphaV), Dark Reading, Aaron Sandeen

https://www.darkreading.com/vulnerabilities-threats/everything-you-need-to-know-about-blackcat-alphav-, September 8, 2022

Dark Web Profile: BlackCat (ALPHV), SOCRadar, https://socradar.io/dark-web-profile-blackcat-alphv/ , August 26, 2022

FBI Releases IOCs Associated with BlackCat/ALPHV Ransomware, Cybersecurity & Infrastructure Security Agency, https://www.cisa.gov/news-events/alerts/2022/04/22/fbi-releases-iocs-associated-blackcatalphv-ransomware, April 22, 2022

BlackCat/ALPHV Ransomware Indicators of Compromise, Federal Bureau of Investigations,

https://www.ic3.gov/Media/News/2022/220420.pdf, April 19, 2022

Alpha Spider, CrowdStrike, https://www.crowdstrike.com/adversaries/alpha-spider/, (n.d.)

Advanced Persistent Threats (APTs), Mandiant, https://www.mandiant.com/resources/insights/apt-groups, (n.d.)

Assembling the Russian Nesting Doll: UNC2452 Merged into APT29, Mandiant, https://www.mandiant.com/resources/blog/unc2452-merged-into-apt29, April 27, 2022

The 10 most dangerous cyber threat actors, CSOOnline, Andrada Fiscutean, https://www.csoonline.com/article/570739/the-10-most-dangerous-cyber-threat-actors.html, May 24, 2021

Ocean Lotus, Council on Foreign Relations, Cyber Operations Home, https://www.cfr.org/cyber-operations/ocean-lotus (n.d.)

Cyber Espionage is Alive and Well: APT32 and the Threat to Global Corporations, Mandiant, Nick Carr, https://www.mandiant.com/resources/blog/cyber-espionage-apt32, November 29, 2022

Groups, The MITRE Corporation, https://attack.mitre.org/groups/, (n.d.)

USB Drives Spread Spyware as China's Mustang Panda APT Goes Global, Dark Reading, Elizabeth Montalbano, https://www.darkreading.com/threat-intelligence/usb-drives-spyware-china-mustang-panda-apt-global, June 22, 2023

Chinese 'Mustang Panda' Hackers Actively Targeting Governments Worldwide, The Hacker News, Ravie Lakshmanan, https://thehackernews.com/2022/11/chinese-mustang-panda-hackers-actively.html, November 19, 2022

2023 Data Breach Investigation Report, Verizon, https://www.verizon.com/business/resources/reports/dbir/ (n.d.)

Cost of a Data Breach Report 2023, IBM, https://www.ibm.com/reports/data-breach, July 2023

Data Breaches Hit Lots More People in 2022, CNET, Bree Fowler, https://www.cnet.com/tech/services-and-software/data-breaches-hit-lots-more-people-in-2022/, January 25, 2023

Identity Theft Resource Center's 2022 Annual Data Breach Report Reveals Near-Record Number of Compromises, Identity Theft Resource Center, https://www.idtheftcenter.org/post/2022-annual-data-breach-report-reveals-near-record-number-compromises/, January 25, 2023

Chapter 4:

CIA Trojan Causes Siberian Gas Pipeline Explosion, The Repository of Industrial Security Incidents, https://www.risidata.com/Database/Detail/cia-trojan-causes-siberian-gas-pipeline explosion, (n.d.)

Reagan Approved Plan to Sabotage Soviets, The Washington Post, David E. Hoffman, https://www.washingtonpost.com/archive/politics/2004/02/27/reagan-approved-plan-to-sabotage-soviets/a9184eff-47fd-402e-beb2-63970851e130/, February 27, 2004

Review of the 1982 Soviet Explosion, Wordpress Website, Georgia Hill, https://blogs.ncl.ac.uk/ghill4/sample-page/review-2/, (n.d.)

Top 10 Cybersecurity Statistics, Worth Insurance, Darren Craft, https://www.worthinsurance.com/post/cybersecurity-statistics, February 16, 2023

Cybersecurity & Infrastructure Security Agency, Secure Our World, https://www.cisa.gov/secure-our-world (n.d.)

How Many Cyber Attacks Per Day: The Latest Stats and Impacts in 2023, Astra, Nivedita James, https://www.getastra.com/blog/security-audit/how-many-cyber-attacks-per-day/, May 2, 2023

10 Types of Social Engineering Attacks, CrowdStrike, https://www.crowdstrike.com/cybersecurity-101/types-of-social-engineering-attacks/, December 28, 2021

10 Most Common Types of Cyber Attacks, CrowdStrike, https://www.crowdstrike.com/cybersecurity-101/cyberattacks/most-common-types-of-cyberattacks/, February 13, 2023

Malvertising, Center for Internet Security, Dilan Samarasinghe, https://www.cisecurity.org/insights/blog/malvertising, (n.d.)

New Cybersecurity Technology, Security Magazine, Maria Henriquez, https://www.securitymagazine.com/articles/98726-new-cybersecurity-technology-2022, December 19, 2022

Five Cybersecurity Technologies Disrupting the Security Landscape, Truefort, Security Research, https://truefort.com/disruptive-cybersecurity-technologies/, February 21, 2023

Chapter 5:

The Untold Story of NotPetya, the Most Devastating Cyberattack in History, Wired, Andy Greenberg, https://www.wired.com/story/notpetya-cyberattack-ukraine-russia-code-crashed-the-world/, August 22, 2018

WannaCry explained: A perfect ransomware storm, CSO Online, Josh Fruhlinger, https://www.csoonline.com/article/563017/wannacry-explained-a-perfect-ransomware-storm.html, August 24, 2022

SolarWinds Cyberattack Demands Significant Federal and Private-Sector Response (infographic), U.S. Government Accountability Office, https://www.gao.gov/blog/solarwinds-cyberattack-demands-significant-federal-and-private-sector-response-infographic, April 22, 2021

Colonial Pipeline hack explained: Everything you need to know, TechTarget, Sean Michael Kerner, https://www.techtarget.com/whatis/feature/Colonial-Pipeline-hack-explained-Everything-you-need-to-know, April 26, 2022

German industrial giant thyssenkrupp targeted in a new cyberattack, Security Affairs, Pierluigi Paganini, https://securityaffairs.com/139870/hacking/thyssenkrupp-targeted-cyberattack.html, December 21, 2022

European banks hit by Russian hackers, Finextra, https://www.finextra.com/newsarticle/42507/european-banks-hit-by-russian-hackers#:~:text=A%20gang%20of%20pro%2DRussian,the%20victim%20of%20an%20attack., June 20, 2023

Cybersecurity 101: The Fundamentals of Cybersecurity, CrowdStrike, https://www.crowdstrike.com/cybersecurity-101/, (n.d.)

Implementing a Zero-Trust Strategy? Start with Universal ZTNA, Network World, Peter Newton, https://www.networkworld.com/article/3674991/implementing-a-zero-trust-strategy-start-with-universal-ztna.html, September 23, 2022

Implementing a Zero Trust Architecture, NIST National Cybersecurity Center of Excellence, https://www.nccoe.nist.gov/projects/implementing-zero-trust-architecture, (n.d.)

Zero Trust Architecture, NIST, NIST Special Publication 800-207, Scott Roe, Oliver Borchert, Stu Mitchell, Sean Connelly, https://nvlpubs.nist.gov/nistpubs/SpecialPublications/NIST.SP.800-207.pdf, August 2020

Artificial intelligence is playing a bigger role in cybersecurity, but the bad guys may benefit the most, CNBC, Bob Violino, https://www.cnbc.com/2022/09/13/ai-has-bigger-role-in-cybersecurity-but-hackers-may-benefit-the-most.html, September 13, 2022

Why AI is the key to cutting-edge cybersecurity, World Economic Forum, Santeri Kangas, https://www.weforum.org/agenda/2022/07/why-ai-is-the-key-to-cutting-edge-cybersecurity/, July 22, 2022

How Blockchain Could Revolutionize Cybersecurity, Forbes, Robert Napoli, https://www.forbes.com/sites/forbestechcouncil/2022/03/04/how-blockchain-could-revolutionize-cybersecurity/?sh=62f9c6a3a41f, March 4, 2022

Quantum computing and cybersecurity: How to capitalize on opportunities and sidestep risks, IBM, Dr. Walid Rjaibi, Dr. Sridhar Muppidi, Mary O'Brien, https://www.ibm.com/thought-leadership/institute-business-value/en-us/report/quantumsecurity, July 18, 2018

ATT&CK, The MITRE Corporation, https://attack.mitre.org/, (n.d.)

Chapter 6:

Six Cybersecurity Trends Influencing CISOs Today, JM Search, Jamey Cummings, https://jmsearch.com/blog/six-cybersecurity-trends-influencing-cisos-today/, August 16, 2021

SEC Adopts Rules on Cybersecurity Risk Management, Strategy, Governance, and Incident Disclosure by Public Companies, U.S. Securities and Exchange Commission, https://www.sec.gov/news/press-release/2023-139, July 26, 2023

The CISO Carousel and Its Effect on Enterprise Cybersecurity, Security Week, CISO Strategy, Kevin Townsend, https://www.securityweek.com/the-ciso-carousel-and-its-effect-on-enterprise-cybersecurity/#:~:text=The%20average%20tenure%20of%20a,meaningful%20seat%20at%20the%20table., September 26, 2023

2023 NACD Director's Handbook on Cyber-Risk Oversight, NACD (National Association of Corporate Directors), NACD and Internet Security Alliance, https://jmsearch.com/articles/six-cybersecurity-trends-influencing-cisos-today/, March 24, 2023

The CISO Job And Its Short Tenure, Forbes, Gary Hayslip, https://www.forbes.com/sites/forbestechcouncil/2020/02/10/the-ciso-job-and-its-short-tenure/?sh=1eb841023482, February 10, 2020

Glossary, NIST (National Institute of Standards and Technology), Information Technology Laboratory, Computer Security Resource Center, https://csrc.nist.gov/glossary/term/cyber_resiliency, (n.d.)

Developing Cyber-Resilient Systems: A Systems Security Engineering Approach, NIST Special Publication 800-160, Volume 2, NIST (National Institute of Standards and Technology), Ron Ross, Victoria Pillitteri, Richard Graubart, Deborah Bodeau, Rosalie McQuaid, https://nvlpubs.nist.gov/nistpubs/SpecialPublications/NIST.SP.800-160v2r1.pdf, December 2021

Life Inside The Perimeter: Understanding the modern CISO, Nominet Cyber Security, https://media.nominet.uk/wp-content/uploads/2019/02/12130924/Nominet-Cyber_CISO-report_FINAL-130219.pdf (n.d.)

5 Ways CISOs Can Deepen Board Relationships, Shared Assessments, Sabine Zimmer, https://sharedassessments.org/blog/5-ways-cisos-can-deepen-board-relationships/, February 12, 2020.

Why Your Board Of Directors Should Focus On Building Your CISO's Self-Resilience, Forbes, Lucia Milica Stacy, https://www.forbes.com/sites/forbestechcouncil/2023/01/27/why-your-board-of-directors-should-focus-on-building-your-cisos-self-resilience/?sh=140a6086293d, January 27, 2023

Key Relationship Strategies CIOs And CISOs Must Master To Win Over The Board, Forbes, Mark Weatherford, https://www.forbes.com/sites/markweatherford/2020/10/19/5-key-board-relationship-strategies-cios-and-cisos/?sh=6f4a3aa87970, October 19, 2020

Guidance for CISOs Pursuing Board Positions, IANS Research, https://www.iansresearch.com/resources/all-blogs/post/security-blog/2023/07/18/guidance-for-cisos-pursuing-board-positions, July 18, 2023

Five Best Practices For CISOs When Speaking To The Board, Forbes, John Maynard, https://www.forbes.com/sites/forbestechcouncil/2022/09/01/five-best-practices-for-cisos-when-speaking-to-the-board/?sh=16290d653d58, September 1, 2022

How CISOs and cybersecurity execs can get board ready, SC Magazine, https://www.scmagazine.com/analysis/how-cisos-and-cybersecurity-execs-can-get-board-ready, June 6, 2023

The Board of Directors Will See You Now, Dark Reading, Marc Gaffan, https://www.darkreading.com/risk/the-board-of-directors-will-see-you-now, March 23, 2023

Chapter 7:

Iran Cyber Threat Overview and Advisories, Cybersecurity & Infrastructure Security Agency, https://www.cisa.gov/topics/cyber-threats-and-advisories/advanced-persistent-threats/iran, (n.d.)

Washington Metropolitan Area Transit Authority, Maps, https://www.wmata.com/schedules/maps/, (n.d.)

Advanced Persistent Threat, CrowdStrike, https://www.crowdstrike.com/cybersecurity-101/advanced-persistent-threat-apt/, February 28, 2023

Oceans and the Law of the Sea, United Nations, https://www.un.org/en/global-issues/oceans-and-the-law-of-the-sea#:~:text=United%20Nations%20Law%20of%20the%20Sea%20Convention%20(UNCLOS)&text=The%20convention%20has%20resolved%20several,up%20to%20200%20miles%20offshore, (n.d.)

Computer Fraud and Abuse Act, The United States Department of Justice, https://www.justice.gov/jm/jm-9-48000-computer-fraud, May 2022

InfraGard Fact Sheet, U.S. Department of Justice, Federal Bureau of Investigation, https://www.infragard.org/Files/InfraGard_Factsheet_2-24-2022.pdf, February 24, 2022

COVID-19, Centers for Disease Control and Prevention, https://www.cdc.gov/coronavirus/2019-ncov/index.html, (n.d.)

Glossary, NIST (National Institute of Standards and Technology), Information Technology Laboratory, Computer Security Resource Center, https://csrc.nist.gov/glossary/term/trojan_horse, (n.d.)

Chapter 8:

ALGORITHMIC WARFARE: NATO Ponders Using Article Five for Cyber Attacks, Josh Luckenbaugh, National Defense Magazine, https://www.nationaldefensemagazine.org/articles/2023/8/31/nato-ponders-using-article-five-for-cyber-attacks, August 31, 2023

The North Atlantic Treaty, North Atlantic Treaty Organization, https://www.nato.int/cps/en/natohq/official_texts_17120.htm, Last updated April 10, 2019

Technology Partnerships: Five Essential Strategies for Success, Forbes, Robert Garrett,https://www.forbes.com/sites/forbesbusinessdevelopmentcouncil/2019/08/16/technology-partnerships-five-essential-strategies-for-success/?sh=4b18bc425ecd, April 16, 2019

What Makes Innovation Partnerships Success, Harvard Business Review, Paola Cecchi-Dimeglio, Taha Masood, Andy Ouderkirk, https://hbr.org/2022/07/what-makes-innovation-partnerships-succeed, July 14, 2022

Emerging cyber threats in 2023 from AI to quantum to data poisoning, CSO Magazine, Mary K. Pratt, https://www.csoonline.com/article/651125/emerging-cyber-threats-in-2023-from-ai-to-quantum-to-data-poisoning.html?utm_date=20230911211035&utm_campaign=CSO%20US%20First%20Look&utm_content=Slot%20One%20Title%3A%20Emerging%20cyber%20threats%20in%202023%20from%20AI%20to%20quantum%20to%20data%20poisoning&utm_term=CSO%20US%20Editorial%20Newsletters&utm_medium=email&utm_source=Adestra&huid=a1e44047-e57d-4020-8698-8f25f68bbdf8, Sep 07, 2023

Conclusion:

What the Russian Invasion Reveals About the Future of Cyber, Jon Bateman, Nick Beecroft, Gavin Wilde, **are**

https://carnegieendowment.org/2022/12/19/what-russian-invasion-reveals-about-future-of-cyber-warfare-pub-88667, December 19, 2022

7 Pressing Cybersecurity Questions Boards Need to Ask, Dr. Keri Pearlson, Nelson Novaes Neto, Harvard Business Review, https://hbr.org/2022/03/7-pressing-cybersecurity-questions-boards-need-to-ask, March 4, 2022

Glossary, NIST CSRC (National Institute of Standards and Technology Computer Security Resource Center), https://csrc.nist.gov/glossary/term/cyber_resiliency

THANK YOU FOR READING MY BOOK!

CLAIM YOUR FREE WELCOME CALL

Just to say thanks for buying and reading my book, I would like to provide a way for you to connect with me, no strings attached!

Simply Scan the QR Code Here:

I appreciate your interest in my book and value your feedback as it helps me improve future versions. I would appreciate it if you could leave your invaluable review on Amazon.com with your feedback. Thank you!